# Evolving Knowledge in Natural Science and Artificial Intelligence

J. E. Tiles
Department of Philosophy, University of Hawaii

G. T. McKee
Department of Computer Science, University of Reading

G. C. Dean
Civil Aviation Authority (RSRE, UK)

**Pitman**

PITMAN PUBLISHING
128 Long Acre, London WC2E 9AN

A Division of Longman Group UK Limited



First published in Great Britain 1990

**British Library Cataloguing in Publication Data**

Evolving knowledge in natural science and artificial
intelligence.
1. Artificial intelligence
I. Tiles, J. II. McKee, G. T. III. Dean, G. C.
006.3

ISBN 0-273-08821-1



ISBN 0 273 08821 1

Printed and bound in Great Britain by
Biddles Ltd, Guildford and King's Lynn

# Contents

# Preface

The present book grew out of an interest in what the philosophy of science might contribute to the research work undertaken under the umbrella of "artificial intelligence" (AI) and how developments in AI might change the way philosophy of science is done. In particular it was the interest of Professor Tom Addis of the Department of Computer Science at the University of Reading, which lead through a chain of contacts to a decision by the British Society for the Philosophy of Science (BSPS) to hold its annual conference at Reading on the specific subject which is the title of this book. Organizers of the conference, both on the committee of the BSPS and at Reading, felt that the papers being brought to the conference could be used to open up a wider discussion – and a possibly fruitful collaboration – between the AI community and those philosophers whose teaching and research center around science. The Introduction argues that the papers illustrate a potentially fruitful dialogue between these disciplines, and explains how the papers may be used to develop an appreciation of the issues and help to initiate the dialogue.

The conference organizers wish to acknowledge generous financial support from Oxford University Press (Publishers of the British Journal for the Philosophy of Science), The International Union of History and Philosophy of Science (Division of Logic, Methodology and Philosophy of Science) and from the Science and Engineering Research Council's "Logic for IT" Initiative. Special acknowledgement must also be made to Greg Hunt of the University of Warwick for his tireless efforts in helping to put the conference on a sound financial footing. Thanks also go to Tom Addis and Dov Gabbay for very useful guidance.

# Chapter 1

# Introduction

Jim Tiles
*Department of Philosophy, University of Hawaii at Manoa, 2530 Dole Street, Honolulu.*

It is inviting to compare the rational development of a scientific discipline with the processes which occur in biological evolution. How far that comparison can profitably be carried is a matter of dispute (see Roger Young below, p. 33, on Paul Thagard's previously published views[1]) but one important difference is this: although a scientific discipline, considered as a social institution, does change (does evolve) as it reproduces itself, a crucial factor which determines the direction of that evolution is the idea which existing practitioners (the present generation) have of their own activity.

This idea may not match existing practice and there may or may not be an awareness of this failure to measure up to the idea. In any case the present generation will convey to its intellectual heirs (post-graduates, post-doctoral fellows, and junior colleagues) not only the habits and techniques which constitute the discipline, but also the conception of what the discipline should be, it shortcomings, its aspirations. This idea will be one of several constituents determining the future development

[1]During the discussion of his own paper Thagard responded to the idea of using genetic algorithms to generate the sort of conceptual change found in science by remarking that the representations which could be treated by such algorithms were as yet insufficiently powerful to provide a vehicle for scientific thought.

of the discipline; it will affect what constitutes an acceptable doctoral project, research proposal, journal article. It will give the members of the discipline a sense of direction.

A discipline, especially a relatively new discipline, which aspires to the status of 'a science' does not generate this idea of its aspirations and present shortcomings purely by reflecting on its own activities. Its practitioners will compare what they do with what they believe happens in other "scientific" disciplines. Of course one has to be careful in making such comparisons that one does not try to import specific methods and techniques which are inappropriate to one's own subject matter. But even if nothing is borrowed more specific than a belief in the importance of experiment and the value of mathematical representations, a demand that phenomena be explained and that those explanations be tested by what they do to our ability to predict and control; conceptions of what ideally a science should be will contribute to the development of any new "science." No new discipline escapes this influence, and there are only two general responses open to someone in such a discipline:

1. Let it happen without examining too closely whether what is done in other sciences is accurate or whether goals are appropriate.

2. Mark carefully what is taken to be better and worse practice and what is taken to happen in other sciences and include the refinement of these conceptions on the research agenda of the discipline.

Following the second alternative will draw one into considering the philosophy of the science which one is working to establish.

If it is accepted that the second alternative cannot reasonably be avoided (or evaded) it does not follow that one should call in the specialists whose professional responsibility is philosophy of science and ask them to set the science on a proper footing, as one might call in an accountant to set financial affairs on a proper footing. It does not follow that a philosopher of science should have a place on the research team. However, it does follow that one should seize the occasional opportunity to air the prevailing ideas of what is being done and what ought to be done in front of someone with a wider experience of such conceptions in order to see how such views might be expanded or refined. As with all professional services one takes a chance on the quality of the practitioner when calling upon his or her expertise, but a good philosopher of science should find any account of first hand experience a fair exchange for what may be learned from him or her. The philosopher encountered may well have first hand experience of another science but a broad experience of other sciences comes – of necessity these days – second hand.

So far this applies to any relatively new scientific study, even of some hitherto neglected part of the non-living natural world. There are features of computer science and the closely related field of artificial intelligence which make the philosophy of science of even more concern than in other disciplines and these will be touched upon shortly. But consider how what has been said so far applies to the scientific study of "intelligent machines." In a recently published article, "What kind of field is AI?" [2] Alan Bundy diagnosed "malaise" in AI research on the basis of the following symptoms: the inability of referees for conferences and journals to agree on criteria for accepting papers, the inability of many AI research projects (programs) to apply to more than a very few very restricted examples, the lack of published accounts of how programs came to be written, the lack of continuity (with a few "honourable exceptions") between research projects.[3]

[2]Bundy's article appears in the Proceedings of the workshop on the Foundations of Artificial Intelligence, New Mexico, 1987. At the conference represented by this volume Bundy appeared jointly with Aaron Sloman and advanced and defended these views in a form which did not differ from their 1987 formulation. The next two paragraphs aim to provide a sufficient summary of Bundy's article to complete the context in which, in particular, Sloman's paper was presented. The editors are grateful to Professor Bundy for correcting a misrepresentation of his position in an earlier draft, and for raising several other points which led to improvement in the final version of this introduction.

[3]As a fringe event at the Reading conference Professor Tom Addis initiated a wide ranging discussion of the present state of AI research with the questions, "Do we have an AI Research Programme?" and "Do we need a (new, different, additional) research programme?" Those inclined to answer either 'yes' or 'no' to the first question seemed to offer their differing answers on the same basis, "Yes, we have several research programmes." "No, we haven't a research programme, we have several." Whether this was viewed positively or negatively appeared to depend on

Of course, not every face flushed with the heat of debate is symptomatic of disease, and the absence of disagreement and of challenges that can only be met by ad hoc devices which defy tidy explanations, may be symptoms of a science which is "finished" – not only in the

---

how much one valued unity in one's discipline. What there seemed to be genuine disagreement over, however, was whether AI counted as a sufficiently mature science to possess one or more "paradigms," – T.S. Kuhn's technical term, for all its reputed ambiguity seemed acceptable currency to the assembled company. If AI lacks paradigms, its state of development would have to be compared to that of the social sciences. If however, Marr's treatment of vision, the use of mathematical logic, or work on informal, non-monotonic logic, counted as paradigms (whether thriving or not), then AI merited comparison with physics or chemistry. How one assesses the situation may depend on how severely one measures AI's accomplishments against the goal implicit in its name. It was observed by those taking a dim view that it was nowhere near producing an artificial idiot, let alone an artificial intelligence. Those taking a more cheerful view seemed on the whole less concerned with producing "an" intelligence ("an autonomous intelligence" was the way Aaron Sloman put it in the discussion) and were content with producing some behaviour which is characteristic of intelligent beings in one or another circumscribed context. Even those who felt that one could not produce intelligence by any sum of such intelligent behaviours were apparently ready to accept that to attempt to do more than this at this juncture might simply be unreasonably ambitious. Thus, although Addis's second question clearly invited discussion of the prospects for a deliberate attempt to manage the diversity of AI research from a unified standpoint (to control from above as it were, even if only to add one more direction of research), there was little enthusiasm for this, even among those who viewed the lack of unity in the discipline as a matter for regret. The consensus seemed to be that any unity would have to emerge out of the diversity of efforts presently being made; unity 'by design' would not be real unity; competition between approaches (paradigms, programmes) is in any case productive. However, a sanguine pluralism which proposes to tolerate all approaches and relies on results to determine whether a line of research has anything to contribute to a discipline, does not adequately face up to the question of how resources (money, equipment, the investment of time by the most promising young research workers) are to be allocated. Historically the "scientific research programmes" which historians and philosophers make the subjects of their studies were not dependent on public funding and programmes, or components of such programmes, did not compete directly with one another for resources. Where resources are finite and hence decisions have to be made, not just with one's own time and money, but with someone else's, an account will be demanded, and there is no way to offer a coherent justification except by reference to a unified conception of what is being done and what might be done instead. Forming such a conception may be premature intellectually, but it may be forced on any new scientific discipline by present circumstances.

sense of "complete" but in the sense of "dead and boring." One is, of course, working to bring one's science to a more finished state, but how soon can one expect to reach it? It is important not to forget, what historians and philosophers of science have recently come to discuss openly, the awkward stages in the development of a science, the contributions which periods of turmoil make to the shaping of solutions, the dangers, which the "great mistakes" illustrate, of reaching too quickly for the kind of solution one is expecting to find.

In order to clear a space where a discipline can flourish which is free of the blemishes which disfigure AI as presently constituted, Bundy looked beyond his own field and drew an analogy with the natural sciences and engineering. The natural sciences of physics and chemistry study and try to model "real world phenomena"; applied engineering generates products with commercial or military value. In between there is scope for an activity of "pure engineering" to generate techniques which do not have immediate application, which are not theoretical (in the sense of being capable of being true or false), but which may be useful to applied engineers or may stimulate the model building activities of natural scientists. "Basic AI" is the pure engineering counterpart of the applied activities of software and hardware design for industry and the natural science of cognitive psychology. The techniques of AI, algorithms, architectures, ways of representing knowledge and ways of eliciting knowledge, are unified by the aim of generating automated devices which "have the potential for simulating intelligent behaviour." Within "Basic AI" we can expect and demand the symptoms of health (which Bundy reached by "generalizing and negating" what he saw as the symptoms of malaise): clear, powerful, simple and complete techniques which indisputably behave as they are required to behave. Commercial and military aims may impose ad hoc devices on applied AI; the attempts of cognitive scientists to match the behaviour of their models to the behaviour of humans may likewise admit of only very approximate and ungeneralizable techniques, but the pure engineering of basic AI should be able to free itself of these defects.

An assessment of how accurate this view is concerning what might take place in other disciplines would require a closer look at those disciplines, but the question of whether this conception is appropriate to AI can be raised within the AI community. Aaron Sloman's contribution to this volume, for example, can be seen as challenging some aspects of Bundy's argument, but one should be careful not to conclude that Sloman would dismiss Bundy's criteria for assessing AI work. "Wherever possible we should use neat [conforming to Bundy's criteria] theories to guide our designs and explanations. In particular, scruffy systems should be based on neat designs."[4](p. 19) Nor is Sloman expressing a preference for working in what Bundy classifies as "applied AI." When Sloman argues for the inevitability of having to work with scruffy systems, he is suggesting that one cannot do any really useful basic AI without tackling real life problems and thereby taking the risk of catching the flu of ad hoc techniques, or without exposing oneself to the pneumonia of endless disputes over how to understand the way real (natural) intelligent systems function. Once it is realized that there is not sufficient space between applied engineering and pure natural science (indeed philosophers of science are increasingly pointing to the way these two become entangled in experimental science) for a germ free environment, one will have to bow to the force of this consideration: the tasks which really intelligent systems and really useful artifacts have to perform are combinatorially too complex for neat systems (easily

[4]As discussion at the conference and subsequent private communications brought out, the issue between the two men centers on what one is prepared to count as an achievement or as scientific progress. Bundy views the achievements of a discipline as residing in its "neat," conceptually tractable theoretical representations. He compares the situation in AI with that in meteorology where the physics is "neat" (something we understand), but its application to a complex real world produces messiness. Sloman would not only regard the messy outcome of such an application as a scientific discovery, but would stress that separating the neat theory from the messy reality, which we hope to understand, is likely to disengage the engine of scientific progress from its source of change. And if recent experience in meteorology, one of the fields from which "chaos-theory" emerged [see below on Petitot-Smith and Morton], is anything to go by, messiness may be a symptom that a neat theory has misrepresented the phenomena in a profound way. At the same time Bundy's stress on the importance of neatness is a salutary reminder of what it is one needs to do to gain by grappling with the complexity of nature. To make a messy representation of a messy situation is to put two things we do not understand where previously we had only one. What the new outlook on meteorology has suggested is that not only have we looked for the wrong sort of simplicity in the principles which govern weather phenomena, we have overlooked a perspective from which patterns can be seen which reduce the apparent messiness of the weather in unexpected ways.

understood, readily generalizable, free of ad hoc devices) to handle.

Sloman's argument is based on the assumption that to qualify as intelligent a system (natural or artificial) must be able (as a necessary condition) to cope with problems which can be represented as solvable by exploring branching paths in a search space. As the number of steps needed to solve a problem increases, the number of possible branches increases very rapidly (producing a "combinatorial explosion"). Representing problems in this way actually constitutes a simplification, for living creatures have far less knowledge and far more options than appear in this sort of representation of their situation, yet in spite of the enormous complexity, living systems have to cope in relatively little real time.[5] The basis of Sloman's argument is given a striking illustration in the problem addressed and the problems left untreated in the work on which Spirtes, Glymour and Scheines report. Even with favourable assumptions about the variables being measured (e.g. the possibility of giving them a time order) the possible causal structures that could be found for twelve variables is in the region of $10^{20}$. There are programs (Glymour reports on TETRAD II and MEASURED.) which, contrary to well established philosophical cautions about inferring causal structure from statistical correlations, can construct causal structure from statistical data. But they do not generate, something which data sometimes generates in human beings, suspicions that not all the relevant variables have been measured, and they do not generate the "initial latent variable models from the data alone", nor do they generate all the statistically equivalent alternative causal models.

Considerations like these lead Sloman to propose that the agenda of basic AI research be framed by the general problem of how living systems – "existing animals, including especially human beings" – cope

[5]Bundy's insistence that the combinatorial explosion should not blunt our appetite for neatness, asks in effect that we respond to this challenge to our understanding as intelligent systems. The issue between Sloman and Bundy might be argued to boil down to one of tactics. Should significant research effort go into developing neat techniques, which are not inspired directly by messy problems, or should we concentrate on refining messy solutions, which are inspired directly by the messy phenomena we hope eventually to understand? A tactical decision here, if it is not to be made on the toss of a coin, must be made by reference to wide experience of what has led to progress or to stagnation in other scientific endeavors.

with a demanding, partly knowable environment. Although Sloman is clearly looking for inspiration from complicated living systems ("especially human beings"), suggestions for ways of confining the combinatorial explosion could be implicit in the mechanisms by which even the simplest living systems reproduce themselves. The evolution of life forms, after all, can be understood as the history of solutions to problems posed by a changing environment. The time scale for trying solutions is determined by the length of a generation and there is, over the population as a whole, a capacity for parallel processing of proposed solutions depending on how prolific are the mechanisms (mutation, cross over, etc.) for generating variation. This is the general idea which Roger Young discusses. Genetic algorithms borrow certain formal features of the way organisms preserve, adapt and build on existing solutions and thereby keep the otherwise exponential growth of solutions to be tried within manageable proportions. Here AI looks sideways at another scientific field, evolutionary biology, not to refine its conception of how and where it stands as a science, but as a source of material for building its theories and techniques.

Young notes the points of contact between his special field of interest and that generated by another special field, also inspired by living systems, viz. "connectionist architecture" or "neural networks" or systems where processing is "parallel and distributive." This extended family of hardware devices (and software representations of their behaviour) was inspired by the observation that the way the elements of the brains of higher animals are connected does not at all appear to be like the way the elements of a serial processing computer are connected. In the best experimental spirit, configurations imitating the way nerves are connected in higher animals were wired or simulated and properties of such systems were explored. The results have been intriguing. For some (but only some) they hold out the promise of a real revolutionary breakthrough; thus there exists a certain tension in the AI community between those who want to devote time or resources developing this general approach and those who argue that it offers either nothing new or nothing which is not too intrinsically "scruffy" (cp. Sloman, p. 17) to be properly understood.

The tension here has the makings of what philosophers of science sometimes (following Kuhn) refer to as a "paradigm clash," and looking critically at the vogue for connectionism, Derek Partridge draws on this vocabulary as he evaluates the movement in the light of his conception of the sort of science AI is (or ought) to be. The central problem, and Partridge is not alone in pointing this out, is that even where the connectionist results are impressive, we have only the most tenuous hold on how they work. That connectionist architecture imitates the observable characteristics of the way nerves are connected does not mean they simulate the processes underlying the phenomena which interest us, for we do not yet have a firm enough hold on what structural features of the brain are relevant to intelligent responses ("The myth of empirical guidance," p. 65.) And even if we became interested in these systems for their own sake, we have theoretical devices which represent them in such a way as to explain and to allow us to intervene to reconfigure and thereby to check our explanations ("A need to mess with the innards," p. 59.) There are, as Partridge notes, a number of not unpromising formal techniques which may make connectionist architecture theoretically tractable. Given the limits on how experimentalist and interventionist we can be with human beings, we may have no alternative but to see how much the perceptually salient features of neuroanatomy can teach us by trying to mimic them in our simulations. The problem is to know how much one should expect from a new idea in a given time scale. There are no precedents in our experience other than what are to be found in the history of new (experimentally driven) paradigms in the natural sciences and that history goes back only about three hundred years. But if questions of appropriation of time and resources need to be answered intelligently, then that experience should not be overlooked.

Advocates of "connectionism" claim their devices offer new insights into how it is possible for information – specifically information about how to proceed given certain inputs – to be represented in a system. When we represent such information to ourselves it is in the form of rules, that is we describe the system's "competence" as a set of conditions, "To X, respond Y." We take it that the way this competence is embodied in electronic serial processing systems is quite straightforward and that we can explain the performance of a system by reference to this embodiment. But analogous mechanisms are not readily observed in living systems, and as they are merely hypothesized, a doubt hangs over the attempt to use such mechanisms to understand the competence, and to explain the performance, of living systems. Connectionists repudiate the claim that the systems which they try to model, as well as the models which they build, are ill-understood, claiming instead that the competence is embodied in their systems in a way which would not be immediately obvious to someone whose mind was formed around serial processing. If they are correct then the horizons of our conception of knowledge representation are in need, if not of changing, at least of broadening. If the two at present rival conceptions could be fused into a single horizon the discourse between different parts of the AI community, which at present has the features of the diatribes of rival political parties, could turn into something like a dialogue between different branches of government. This is the task which Martin Davies undertakes as he explores and evaluates the idea of "two-tier architecture," specifically the idea of connectionist components organized under an overseer system which is intelligible along classical serial lines.

Connectionism illustrates the way an idea which is inspired by observable physiological features found well down the hierarchy of living creatures (in any animal with a nervous system) can challenge the way we think of the knowledge possessed by the most complex systems (ourselves.) If the project which Davies is pursuing bears fruit, it will ease the pressure to replace wholesale our conception of the way knowledge is represented in systems which respond at all intelligently. We may have to expand that conception and re-interpret some of what we think we know, but there may not have to be a radical break with existing conceptions. Many of these existing conceptions draw inspiration not from features shared by a wide variety of living systems but specifically from what is taken to be the way human beings acquire knowledge to make intelligent responses and the means they have for representing what they have acquired. Now the paradigms of our conception of what is involved when humans acquire, store, process and retrieve information are to be found in science, and the conceptions we possess of scientific activity are products' of the reflective activity which constitutes the philosophy of science. Here is a respect in which philosophy of science impinges on AI in a way it does not touch other scientific disciplines. Science is a prime example of the behavior AI would like its devices to simulate and emulate. Regardless of whether it is done by a scientist or by a philosopher, it is a philosophical activity to produce and refine conceptions of what it is that AI wishes to simulate here.

Reasoning about matters where one cannot reach certainty has, for example, been taken by many philosophers of science as what characterizes modern empirical science and sets it apart from older classical ways of doing science, which disregarded any reasoning that would not yield apodeictic certainty. Thus it is not at all far fetched to see the papers of Havránek and of Paris and Vencovska as developing a field first explored by philosophers of science who had a strong analytic and empirical orientation. Havránek's paper illustrates well the use which Czech logicians have made of the tools of abstract algebra in this field, while Paris and Vencovska seek to devise new formal representations which offer a psychologically more (if not completely) plausible account of how human experts make judgements which have a significant degree of uncertainty. In common with some of those who made earlier attempts to develop, or depart from, models of reasoning and belief formation which are based on probability calculi, the two papers presented here share the conviction that non-deductive as well as deductive reasoning is a mechanically implementable procedure. What sets present efforts apart from past efforts based on this conviction is the availability of machines which could conceivability realize the implementation and not merely save mental effort but preserve and disseminate human expertise and allow that expertise to transcend psychological limits on human memory and concentration. The better we understand our own thinking, the better we will be able to enhance its power.

The candid account of problems still not tractable, with which both Havranek and Paris and Vencovska end, are testimony to pressures which Sloman identified as pushing AI inexorably into working with "scruffy" solutions. And yet human science, which AI would like to simulate, does not consist entirely of a tissue of well-tried guesses. Occasionally a field where our "knowledge" has hitherto consisted only of tacitly formed diagnostic techniques, and the results of applying the methods of epidemiology, is transformed into an area where theory both anchors and guides our judgements. Sloman's paper contains a discussion and an illustration of a key human ability. Confronted by problems which would appear soluble only by a lucky find in a vast search space, humans have the ability to transform the space into one tightly confined and circumscribed. It is easy to overlook this phenomenon of problem space transformation (cp. Thagard, p. 211 on "conceptual [re]organization" vs. "belief revision.") Attempts, already classic, to devise programs with the capacity to learn in the way a scientist learns from nature, such as the BACON programs examined here by Peter Gibbins, simply were not undertaken with enough appreciation of how much of the solution is contained in the way the problem is set up. Although the BACON programs learn in a sense, their power to learn is "circumscribed by the anticipations of the programmer" (p. 173). As already remarked, more recent programs written in the same spirit as the BACON program, such a TETRAD II and MEASURED, reported on here by Spirtes, Glymour and Scheines, do not make the move which commonly follows a restructuring of the problem space, viz. suggesting that different variables need to be measured (p. 198). And they do not generate the "initial latent variable models from the data alone," nor do they generate all the causal models "statistically equivalent to a given causal model (p. 198)". But is it not a contradiction to ask for a program, which by its very nature is a response to a pre-structured problem space, to represent the stage prior to the writing of the program, the structuring of its problem space?[6] This is what appears to be demanded if our artifacts which simulate intelligence are to generate anything like real science which consists, after all, in "products of our language, our history, and of the way we as organisms and agents interact with the world" (Gibbins, p. 175).

Sloman's discussion invites the inference that our ability to restructure a problem space is a function of the connectionist architecture

[6] My fellow editor, Garfield Dean, put this to me, "Your question here is rhetorical and invites a negative response. Currently this would not be unreasonable. However, is there not the possibility that problem spaces can have the domains of other problem spaces as their own domains, even of problem spaces that represent themselves - e.g. arithmetization for the proof of Gödel's theorem? In other words, there is the possibility of completely modelling the whole scientific process including the component concerned with the philosophy of science, i.e. the meta-scientific activity. Perhaps I am writing too much from the view of someone who believes the whole of science can be automated. I realise that you are treading a line between the requirements of AI (optimism) and Philosophy (analysis), but for me the implication of your question was too pessimistic."

of our nervous system; for some neural nets transform certain problem spaces in this way. But so far we understand how neural nets do this, in the restricted class of cases where they work, no better than we understand our own abilities. Now this phenomenon appears at its most striking in cases in the history of science where "revolutionary conceptual change" has taken place. Paul Thagard's article draws on the literature of the philosophy of science as well as learning theory and computer science to explore ways this process might be modeled. To model the process requires at least a general understanding of the phenomena, and Thagard advances six theses characterizing the phenomena. Efforts to reach such a general understanding also need a focus and that focus has been made sharper for Thagard by fixing on the goal of seeing how far, and indeed what aspects of, our ability to transform problem spaces can be implemented artificially.

Our appreciation of the crucial phenomena can be broadened by comparing efforts to model different but importantly related phenomena. The work, on which Eugene Charniak reports, began with the idea that a close analogy of one crucial form of inference, 'abductive' (cp. Thagard's Thesis 5, p. 210), was also found in the comprehension of written texts and so programs were devised which modeled this activity. Charniak's discussion takes us to the next problem which this work should confront: the very presence of a detail mentioned in a written text is an indication that it possesses significance. "Story readers" are assisted by what structures their problem spaces in a way that "plan recognizers" (who have to find in the environment the means to some end) are not. Plan recognition and the discovery (especially of novel) means to ends is closely analogous to explanatory reasoning and the discovery of novel ways to structure a problem space.

One great obstacle to developing a feeling for the phenomenon of problem space restructuring is the way successful solutions and productive structurings of a problem space come to seem part of the very fabric of nature. It becomes psychologically very difficult to treat seriously the possibility that there might ever be more fruitful ways to organize our concepts and thereby our problems and their possible solutions. The same barriers arise when considering the past. Science, as it was conducted before some breakthrough, comes to seem not only quaint but benighted to the point where we think of it as pre-science. Breaking down the barriers that render the past obscure in this way is at least in some cases easier than peering around the barriers that occlude the future. For in some cases we have documentation on episodes in the history of science, which can stimulate the imagination and help us to think our way into a different space of problems and their solutions.

Reading history is not by itself enough. To make history live one has to resurrect options long dead and buried and try to see what it would be for them to be live options. One way, the historian's, is to try to live imaginatively in Aristotle's problem space; another way, which Barry Smith and Jean Petitot illustrate, is to try to make some of Aristotle's ideas do work for us again in some region of our own problem space. The Petitot-Smith project has, moreover, a potential spin-off, for not only does it offer a feeling for possibilities at present not seen, it promises to generate a more "natural" framework for modeling the interaction between human organism and the world that gives rise to modern science. Aristotle, it is often said, represents a stage in the history of science in which common sense is made more precise and rigourous. The next stage of science involved a sharp break with common sense, but it might be premature to try artificially implementing latter stages without understanding how the earlier stage arose.

Adam Morton's useful discussion of the Petitot-Smith project not only draws out further the potential interest of the project for AI, it reminds us of where we can find very recent experiences of the sort of conceptual change that marks the activity of intelligence at its most interesting and challenging: the new mathematics of turbulence, known as "chaos-theory" or "chaology." Morton concludes by discussing the tension which exists between our ordinary ways of looking at mental phenomena and the viewpoint on these phenomena which AI appears to impose and the possibility that we may need a very different way of structuring our problems in order to make any serious advance in understanding the phenomena of cognitive psychology.

The Petitot-Smith project is accurately identified by Morton as belonging to a tradition founded by Kant. Kant is also the ancestor of the idea of a problem space having a structure which it is our responsibility for imposing. There is, as Morton observes, a congenital susceptibility to "metaphysical extravagance" shared by the intellectual descendants of Kantian ideas. This susceptibility is the opposite of that created by the experience of ideas with a long history of success. Instead of mistaking our organization for the furniture of the world, we regard the whole world as something we can structure and restructure at will. As AI research becomes more ambitious and adventuresome, more and more of its practitioners will find the need to strike a sane balance between these extremes. Here AI will find itself drawn into more general philosophy; for the resources for coping with that tension, such as we have, are kept and developed by the professional academics who refer to their subject as "philosophy." The alternative to being drawn into philosophy here is to have to repeat its history – and that history is a long one.

The foregoing remarks were addressed to computer scientists and those engaged in AI research. They attempted to spell out how the philosophy of science makes contact with their interests and how the papers in this book can place them in a position to establish that contact for themselves. Philosophers of science should have no trouble reading the implications in the other direction. A new "scientific discipline" is another field, the location of which elaborates the very idea of science, and whose problems establishing itself illuminate similar problems in other disciplines. That AI should find its subject matter so close to that of the philosophy of science itself may, however, inspire disquiet in some. It may, in that case, be profitable to consider this in the light of the interest in experimental science which has recently become prominent in the philosophy of science.

Experimental science arose when the class of natural phenomena which interests scientists (natural philosophers) was expanded to include human artifacts – artifacts designed specifically to make manifest the principles by which natural phenomena operate. As programs like BACON are replaced by conceptually more sophisticated artifacts which "have the potential for simulating intelligent behaviour," psychology which has existed as a laboratory science for little more than a century, will move more fully into a new laboratory phase. And it is possible that more and more claims about the nature of scientific thinking – like the vulgar empiricism which was embodied in the BACON programs – will become what amount to experimental subjects. Thus the philosophy of science, which has studied certain forms of human cognitive activity from normative and historical standpoints will have to determine for itself whether it should acquire a laboratory dimension. And before this takes place, instead of observing and participating in the self-examinations of various scientific disciplines, the philosophy of science may be called upon to engage in its own self-examination.

# Chapter 2

# Must Intelligent Systems Be Scruffy?

Aaron Sloman
*School Of Cognitive and Computing Sciences, University of Sussex, Brighton*

## 2.1 Introduction: Neats vs Scruffies

You would probably be very unhappy about air travel if you were told that landing was always under the control of programs that were so complex and messy that nobody really understood how they worked, and nobody had ever proved that they would do something sensible in all the situations that could possibly arise. Would you be equally unhappy to learn that your life was in the hands (excuse the metaphor) of a human brain?

There has been a long-standing opposition within AI between "neats" and "scruffies" (I think the terms were first invented in the late 70s by Bob Abelson at Yale University).

The neats regard it as a disgrace that many AI programs are complex, ill-structured, and so hard to understand that it is not possible to explain or predict their behaviour, let alone prove that they do what they are intended to do. John McCarthy in a televised debate in 1972 once complained about the "Look Ma no hands!" approach. Similarly,

Carl Hewitt, complained around the same time, in seminars, about the "Hairy kludge (pronounced klooge) a month" approach to software development. (His "actor" system was going to be a partial solution to this.)

The scruffies regard messy complexity as inevitable in intelligent systems and point to the failure so far of all attempts to find workable clear and general mechanisms, or mathematical solutions to any important AI problems. There are nice ideas in the General Problem Solver, logical theorem provers, and suchlike but when confronted with non-toy problems they normally get bogged down in combinatorial explosions. Messy complexity, according to scruffies, lies in the nature of problem domains (e.g. our physical environment) and only by using large numbers of ad-hoc special-purpose rules or heuristics, and specially tailored representational devices can problems be solved in a reasonable time.

Moreover, the scruffies tend to complain that the neats spend so much time worrying about form (e.g. defining formal syntax and semantics for their notations) that they ignore content (e.g. what information about the world an intelligent system really needs). They sometimes even liken the "theorem envy" of those who hanker after mathematical rigour to the kind of arid statistics-based research in psychology and the social sciences that uses sophisticated mathematical techniques (and much computer power) with little or no increase in understanding of any interesting phenomenon.

Recently the conflict has taken on a new twist, with the rise of connectionism. Here we have a branch of AI (yes, it is part of AI, not a new rival discipline), that is heavily mathematics-based, yet, although the general principles on which a particular network learns during its training period may be well understood, the operation of the final system when applied to real tasks generally depends on a totally opaque network of connections between processing units and weights that modify their influence on one another. Only in relatively simple cases is it possible to interrogate the "hidden layers" of a neural net and find out what they are doing or why it works. So it looks increasingly as if systems based on neural nets will, after training, be even more inscrutable than AI programs: we'll just have to trust that it is safe to extrapolate from the test cases where they appear to perform well – exactly as we do with people to whom we give responsibility. Alas!

This is not unlike the situation in weather forecasting, where the general physical principles underlying the weather are well understood but the boundary conditions are impossible to measure and represent with complete accuracy and their large scale effects are impossible to compute within acceptable margins of error: small errors in initial inputs to equations can produce unpredictably large errors in their solutions.

Anyhow, quite apart from the implications of connectionist models of neural nets there are more general things to be said about the neat/scruffy conflict. I shall try to separate different strands in the conflict and will conclude that there are good reasons why intelligent systems performing in real time in a real world, or even in some purely formal problem domains, will have to be scruffy. Still, wherever possible we should use neat theories to guide our designs and explanations. In particular, scruffy systems should be based on neat designs.

We can contrast the theories about the world and how to relate to it that are used by working intelligent systems with theories about the design of such systems. I'll call the former object-level-theories and the latter meta-level-theories. My point is that we should strive to make meta-level-theories as neat as possible, including the theories that explain why object-level-theories are bound to be scruffy.

## 2.2 The Scope of AI

One source of optimism about the possibility of neat designs is an excessively narrow view of the scope of AI. For example, there are some who tend to think of AI as synonymous with the field of expert systems, and it is true that for many (though not all) expert systems the problem domain is sufficiently restricted that complete analysis is possible. I prefer a definition that reflects the range of work reported at AI conferences, in AI journals, and the interests and activities of some of the leading practitioners, including founders of the subject. From this viewpoint AI is a very general investigation of the nature of intelligence and the principles and mechanisms required for understanding or replicating it, and this includes topics like vision and motor control which are important for complete intelligent systems embodied in the physical world.

Like all scientific disciplines AI thus construed has three main types of goal, empirical, theoretical, and practical, encompassing:

1. Empirical study and modelling of existing intelligent systems (including human beings and other animals);

2. Theoretical analysis and exploration of possible intelligent systems and possible mechanisms, architectures or representations usable by such systems, and using the results to explain the empirical phenomena observed under (1);

3. Solving practical problems in the light of (1) and (2), namely:

    (3a) Using the understanding gained in (2) to deal with problems of existing intelligent systems (e.g. problems of human learning or emotional difficulties),

    (3b) Using the theoretical understanding to design new useful intelligent machines,

    (3c) Using the theoretical understanding to design new machines that may not in themselves be intelligent but nevertheless usefully complement human intelligence.

In the course of these activities AI generates new sub-problems, and these lead to new concepts, new formalisms, and new techniques. On this general definition there is little difference between AI and Cognitive Science, though in fact many of the people who call themselves cognitive scientists are ignorant of the contents or relevance of AI to their work.

Much of the theoretical work that falls under this general view of AI includes understanding design trade-offs relevant to intelligent systems. For examples there are trade-offs between efficiency and generality, between complexity of design and the ability to degrade gracefully when things go wrong. The rest of this paper is concerned with trade-offs that confront an intelligent system that has to solve problems within constraints of time and memory.

## 2.3 Bow to the Inevitable: Why Scruffiness is Unavoidable

The main reason why scruffiness is inevitable in intelligent systems can be summarised thus: in complex domains, neatness is defeated by combinatorial explosions, unless you have an infinitely fast processor or all eternity to wait for solutions.

The argument runs as follows: except in limited cases (to be discussed below) problem domains have a structure that requires solutions to be found by exploration of branching paths in a search space. The number of possible branches will be (at least) an exponential function of the depth of path to solution. E.g. if N steps are required and the number of possible branches per step is as low as 2, the number of possible paths will be $2^N$. So even if only 50 steps are required, and they can be explored at the rate of a million a second they will need about 35 years for an exhaustive exploration. A mere 70 steps would require about 37 million years. That indicates how exponential functions explode: adding a mere 20 of whatever the cost is an exponential function of multiplies the time or space requirements by about a million for a branching factor of 2 or over 3,000,000,000 for a branching factor of 3. These functions grow so fast that they are not going to be helped much by parallel implementations.

For problem domains that have this character it is very often the case that neat, general, sound and complete algorithms work only on small problems: they do not scale up because they have to do so much detailed analysis of cases that they explode combinatorially.

## 2.4 Non-Explosive Domains

Of course, not all search problems are like that. If you have a very large array of names alphabetically ordered and want to search for a particular name, you can (given suitable addressing mechanisms) use the "binary chop" method, i.e. look at the middle, then decide whether to go left or right, then look in the middle of the remaining region, etc. With this method the relation between problem size and solution time is the reverse of exponential, it's logarithmic. I.e. the size of the problem is an exponential function of the time it takes to solve it rather than the other way round. So even if the problem (the list of names) is made a million times longer, you will typically need only about 20 more steps to find a name. That's how telephone directories work.

There are other problems that have tractable complexity. For example if you are trying to find some combination of N values of variables that solve a problem, instead of searching through all N factorial possible combinations you may be able to structure the problem as a search in a relatively homogeneous N-dimensional vector space, where there is at any point a function that tells you in which direction to move towards the solution point. Then you can home in on the solution fairly directly by moving along the line of steepest ascent to the goal, e.g. using relaxation techniques.

For more complex problems that can be represented in an N dimensional vector space, for instance pattern recognition problems, it is well known that this simple hill-climbing strategy won't work, e.g. because you can end up on a local hillock or get lost rambling aimlessly on a plateau with no clear direction of steepest ascent. However, some kinds of neural nets are capable of being trained on a particular class of problems so that they transform such a space into a different space in which solutions are found very quickly by going along the direction of steepest ascent or descent, though there is still much to be learnt about the classes of problems for which this works, and under what conditions the training methods will produce complete and sound problem solvers.

Unfortunately, it is far from obvious that all the problems that an intelligent agent needs to deal with can be expressed in a form that is amenable to these solution methods. For example for some problems (e.g. searching for proofs in pure number theory) the structure of the domain makes it impossible either to set bounds to the length of proofs or to avoid branching searches for proofs. This makes it impossible to map the search space into a finite dimensional vector space amenable to relaxation or neural net techniques that home in on solutions without combinatorial searching: we are again confronted by these horrendous exponential functions.

This is why nobody tries to solve problems in arithmetic by simply remembering Peano's five axioms and searching for a logical proof based on those axioms. Instead we memorise all kinds of strictly redundant facts derivable from those axioms, including multiplication tables and lots of rules of thumb for solving particular problems (e.g. you can tell if an integer is divisible by five by looking at its last digit in the decimal representation, or whether it is divisible by three by seeing whether the sum of its decimal digits is divisible by three, etc.).

These stored "lemmas" (previously proved results) are facts that are strictly redundant but have been found to be useful for classes of problems that occur frequently. For other problems combinatorial search may still be required.

Of course, if you depend so much on stored facts you have a new problem which is to find a way of organising them so that you can tell which one, if any, is relevant to a particular task. There is no infallible way of doing that: discovering relevance can be has hard a problem as any.

So, for the full potentially infinite collection of arithmetical problems, we are about as badly off with the stored intermediate results as we are without them. Fortunately, however, it is an empirical fact that for most people's lives there's only a comparatively small, bounded, variety of arithmetical problems that they ever have to solve, so it pays to learn the ad-hoc collection of facts and tricks that enable such problems to be solved (with or without our understanding why).

Those unfortunates who do research in pure number theory have no such complete problem solving kit, and therefore have to spend their lifetime groping around without such effective aids to rapid solutions, though even they tend to specialise in sub-fields where they can build up expertise in the form of an ad-hoc collection of useful heuristics, lemmas, and patterns, for deciding what's worth trying next.

In chess, where, unlike the set of possible proofs in number theory, the move tree is finite, it is nevertheless so large that as regards strategies for deciding in a reasonable time where to move, the tree might as well be infinite. Of course there are some important relatively small subtrees for which winning algorithms have been discovered, and good chess players learn patterns on the board to recognize when they have entered such subtrees, or when they are close to entering them.

So unless someone invents an infinitely fast, infinitely large, processing system, any intelligent agent will have to find short cuts to solving problems, even when dealing with formally specified domains. In fact, short cuts are the key to intelligence: a major component of intelligence is *productive laziness.*

Here's an example of a short cut using a special purpose representation. Suppose you are asked to find all possible ways of selecting values between 1 and 9 inclusive for the variables $a$ to $i$, satisfying the following conditions:

1. No two variables have the same value.

2. All the following are true:

$$\begin{array}{ccccccc} a & + & b & + & c & = & 15 \\ a & + & d & + & g & = & 15 \\ a & + & e & + & i & = & 15 \\ b & + & e & + & h & = & 15 \\ c & + & f & + & i & = & 15 \\ d & + & e & + & f & = & 15 \\ g & + & e & + & c & = & 15 \\ g & + & h & + & i & = & 15 \end{array}$$

You can search through all possible ways of assigning the 9 integers to the 9 letters, i.e 9 factorial or 362,880 combinations. However, a little representational and conceptual creativity can almost completely eliminate the searching required. The trick requires the following observations:

(a) The problem is equivalent to mapping the integers 1 to 9 onto locations in a 3 by 3 array of squares labelled thus $\begin{array}{ccc} a & b & c \\ d & e & f \\ g & h & i \end{array}$ such that all rows, columns and diagonals add up to 15.

(b) There are three distinct types of locations in the array, occurring in different numbers of collinear triples:

- the centre ($e$) occurring in four triples,
- corners ($a$, $c$, $g$, $i$) occurring in three,
- middle edges ($b$, $d$, $f$, $h$) occurring in two triples.

(c) So there must be three different types of integers between 1 and 9 such that some occur in four, some in three, and some in two triples adding up to 15.

It takes very little work to divide the nine integers into these three categories. After that assigning them to locations in the array so as to satisfy the conditions requires almost no backtrack searching. After one solution is found all the others are generated quickly by using the symmetries that are blindingly obvious in the geometrical representation though not at all obvious in the original formulation of the problem.

This example of productive laziness used the ability to notice a relationship between an arithmetical problem and a geometric structure, as a result of which some additional arithmetical concepts were created (concepts of different kinds of numbers) which can then be used to constrain the assignment of numbers to letters in such a way as to more or less eliminate searching.

This toy example is typical of creative problem solving with much harder problems in mathematics and elsewhere. Often a problem looks either unsolvable, or solvable only by very tedious exhaustive search, until a relationship between domains, or between problems, is noticed that provides a short cut to a solution.

Sometimes discovering such relationships is just a matter of luck. Sometimes it results from a generalisation or modification of a similar relationship that has already been shown to have heuristic power.

However, for many kinds of problems short cuts work if you are lucky but they are not perfect. Moreover, most good short cuts cannot be worked out in advance on the basis of prior analysis of the domain: they have to be discovered by exploration and analysis, like the way in which a great chess player (or mathematician) learns patterns by observing and generalising what turns up in actual experience instead of trying to do everything in advance by applying general inference procedures to the rules of chess (or the axioms of a branch of mathematics).

So even in formal and mathematical domains there are not going to be neat generally applicable powerful methods that can be used to solve all problems in a reasonable time.

## 2.5 The Physical (Biological, Social) World is Even Harder to Deal With

The requirement to discover rules of thumb and short-cuts by exploration and experience is even more pressing in connection with problem solving in the real world (e.g. finding good ways to build houses that are resistant to wind, rain, cold, etc.). There are several reasons why things are harder in the real world, including the following:

(a) Unlike a formal system like chess or number theory, no complete body of information is available as a starting point: knowledge is always limited. (In fact we are generally both ignorant about some things and misinformed about others.)

(b) The range of possible things to do, and therefore the branching factor in search spaces will be considerably higher than in chess or arithmetic: you can scratch your ear, shout for help, look for a library, try to climb a tree, think of a number, try to remember whether you have been here before, etc., etc. So search spaces are unmanageably large, especially if the task is to find all the solutions, or to find the best one.

(c) Things happen at speed in the world, and very often decisions and actions are required by a deadline in the immediate future. This rules out extensive combinatorial searches.

For all these reasons rapidly accessible, rapidly executable rules of thumb have to be derived or learnt by trial and error. Because the world is treacherous and we are not gods, such rules are highly fallible.

In both the real world and formal domains mistakes are possible. A heuristic induced from actual experience can be erroneous because the wrong generalisation was made, even in a mathematical problem domain. Additional problems in the physical world are poor observation or measurements, or a poor set of concepts for dealing with the phenomena - like trying to think about movement of masses without the concept of acceleration. There are no general, provably correct, algorithms for finding powerful new notations and using them to define new concepts. Scientific and cultural developments are slow, erratic, and often painfully wrong. And this is inevitable.

## 2.6 Limits of Consistency in Intelligent Systems

Given that some of your rules, generalizations and facts may be wrong, one possible (though not infallible) way of detecting that errors exist might be to check whether they are all mutually consistent. If not, something must be wrong, though consistency does not imply that everything is fine.

Alas, consistency checking is itself subject to combinatorial explosions. Even if all complexity in the knowledge base is due to the use of propositional connectives, and no quantifiers are involved, if there are N atomic propositions a consistency check can take up to $2^N$ tests, which will be an impossibly large number for a realistic database. Of course, particular cases can be dealt with more quickly, but in general checking consistency in a large knowledge base is an intractable problem.

Moreover, even if you've managed to achieve internal consistency, maintaining it is a new problem if your beliefs are error prone, and you frequently have to modify beliefs. Because the knowledge base has to be redundant for reasons mentioned previously, retracting a belief raises the problem of finding and dealing with other redundant beliefs, plans, rules, etc. that were derived from the one that has been rejected.

You could in principle embed your knowledge base in a reason maintenance system that kept records of all such dependencies, telling you what else must be retracted as a consequence of correcting errors, but again in practice this strategy will be defeated by the sheer volume of dependency relations, in any realistic knowledge base.

So the only *practical* strategy for revising a belief is to use whatever heuristics are available for finding closely related beliefs that also need to be revised, and then hope that if there are any other beliefs that were derived from the erroneous one you will later independently detect their wrongness and correct them before they lead you into trouble. In practice we don't always avoid such trouble, and the arguments presented here suggest that that is not just because human beings are badly designed but because the problem will inhere in any resource-limited system.

## 2.7 Scruffy Semantics

There is an additional kind of scruffiness that I think is inevitable in intelligent systems inhabiting a rich, and only partly knowable, world: namely scruffy semantics.

Logicians like to define formalisms whose syntax and semantics are specified precisely.

A definition of the syntax of a formalism starts by assuming some class of structures (e.g. all possible sequences of characters in some alphabet, or all possible networks made of certain kinds of nodes and links) and then giving a rigorous definition of a subset of such structures that are taken to be syntactically well formed (in the formalism being defined). Any structure will either be definitely well formed or not.

Human beings often deploy representational formalisms that are open-ended in ways that are incompatible with this conception of syntax. For example, maps are widely used, but there is no set of rules specifying exactly what is and what is not a legal map. It is up to someone creating a map to decide what will be an effective device for the intended purpose, assuming that the eventual user will be creative and intelligent. As a result there is considerable variation in road maps published by different firms. Some of them use considerable ingenuity in inventing new encodings of information, for example whether a link between a motorway and a lesser road allows both entry to and exit from the motorway, or only one of these – a problem that used not to arise for more primitive road systems.

Sometimes the conventions used in a map are explicitly specified in a key, but sometimes it is up to the reader to infer them, using general knowledge of the domain and the purposes of the map.

This kind of open-endedness applies to all kinds of pictures, diagrams, charts, tables, models, etc. used in human communication. The prior existence of some agreed definition precisely specifying a syntax is the exception rather than the rule.

Moreover, if we were not allowed this syntactic creativity in our communicative devices we would be impoverished not only culturally but also scientifically and technically. Even mathematicians often create new diagrams and notations both when searching for proofs and when communicating concepts and proofs to others.

Of course, natural languages are equally prone to creative extension. When a toddler once said to his parents "Today might be much more hotter than it usually be's", in a discussion of whether to go on a picnic, they understood him perfectly. More generally, the syntax of natural languages is constantly evolving, and evolving simultaneously in different directions in different sub-cultures.

Semantic scruffiness is an even deeper issue than syntactic scruffiness. A formal semantics of the kind defined by Tarski [Tarski 1956] maps well formed expressions in a formalism onto elements or subsets of a previously specified set of objects in such a way that whether something is or is not so mapped is determined by a collection of recursive rules.

In real life we are not in a position to specify precisely which sets of objects we are dealing with, or the mapping rules.

Instead we muddle on with a collection of relatively ill-specified notations, and employ structures using such notations with a collection of ill-specified semantic rules. In particular it is not generally the case that for human language the question whether a situation is or is not correctly described by a sentence has a determinate answer. "That is a giraffe" might be indeterminate when a new kind of animal is discovered that has some of the features previously regarded as essential to being a giraffe, but not all of them. Many scientific advances depend on the introduction of new concepts that are later found to have semantic indeterminateness, for instance the concept of mass in Newtonian mechanics. When the indeterminateness is discovered, e.g. because an unanticipated situation arises in which defining criteria generate conflicts, this often catalyses conceptual changes that enable us to extend our concepts and theories to provide us with a deeper more powerful understanding of the world.

A full discussion of this issue is impossible in the space available here. I'll simply baldly assert that for creative intelligent agents groping towards ever increasing knowledge and understanding of an indefinitely rich and complex world it is inevitable that the notations will have both syntactic and semantic indeterminacy of a kind that both limits the applicability of logical and mathematical methods but also provides a stimulus to creative advance.

## 2.8 So Various Kinds of Scruffiness are Inevitable

If all this is correct, then over the years any intelligent learner in the real world, as opposed to some ideal world in the mind of wishful logicians, will build up an ever increasing store of ill-defined information, much of it inevitably redundant, much of it inevitably wrong, and with no hope of keeping it all internally consistent. I.e. real intelligent agents will be irretrievably scruffy.

I have argued elsewhere [Sloman 1987] that there will be additional kinds of scruffiness to do with the requirements of control systems in an agent that has multiple independent sources of motivation that can be triggered by new events in the world, and has limited information and limited time for dealing with most of the problems. A conclusion of that argument is that the kinds of mechanisms that sometimes generate emotional states are not design defects but inevitable consequences of optimising the design, subject to all the problems and constraints of real life.

## 2.9 What Should AI do About This?

One reaction to all this is horrified rejection and a frenzied search for some general way of using formal, mathematical methods to avoid the problems. An alternative reaction is to recognize that a major task for AI is to find ways of designing intelligent agents that reduce, even though they cannot eliminate, the bad consequences of all this scruffiness. This is compatible with trying to identify classes of sub-problems that are completely amenable to rigorous mathematical treatment.

Unfortunately, because we don't know enough about the world to prove that any particular way of minimizing the consequences is optimal, and I suspect that there is NO practically implementable strategy that will be optimal for ALL situations, it seems to me that our best hope is to see what has emerged as a result of millions of years of empirical exploration of the space of possible designs. I.e. we should study existing animals, including especially human beings, to find out how they cope reasonably well in a rich, partly knowable, partly friendly, fast changing environment.

It is unlikely that they do this by proving theorems, even theorems in non-monotonic logics.

Of course, I am not arguing that mathematics and logic are irrelevant to AI. The whole history of science has shown that on the contrary a mathematical approach is often essential for understanding a complex system, and for many aspects of AI it is clear that mathematical analyses of problems and techniques play an essential role both in understanding the nature of the problem to be solved and in designing and evaluating representations and algorithms. My point is only that precisely defined representations and rigorous fully analysed methods have to be restricted to those sub-problems where they are applicable and useful (e.g. aspects of low level vision, design and analysis of various forms of neural computation, some aspects of motor control, complexity analysis, and so on) and that for many features of the design of intelligent systems the *working* system will have to violate canons of mathematical and logical acceptability, but for good, principled, reasons.

## 2.10 Conclusion

Although it may be impossible to build truly intelligent systems that never become scruffy as they struggle to understand the world and how to deal with it, at least we have a neat theory as to why it's impossible.

### References

Sloman, A., [1987]: "Motives Mechanisms Emotions", *Emotion and Cognition*, vol. 1, no. 3, pp.217-234. Reprinted in M. A. Boden (ed.)

*The Philosophy of Artificial Intelligence*, "Oxford Readings in Philosophy" Series, Oxford University Press, to appear 1990.

Tarski, [1951]: "The Concept of Truth in Formalized Languages" in his book *Logic, Semantics, Metamathematics*, New York.

# Chapter 3

# Genetic Algorithms and Scientific Method

Roger A. Young
*Philosophy Department, University of Dundee, Dundee, Scotland*

## 3.1 Introduction to Genetic Algorithms

Genetic algorithms are algorithms based upon principles drawn from the theory of natural selection, especially the theory of natural selection operating together with sexual reproduction. They are used for problems of search, optimization and machine learning. They make some use of randomization, but this is not their distinctive feature. Many other algorithms, e.g. simulated annealing [Davis, 1987, p.8-11],various connectionist models [Rummelhart, 1987, p.151 etc.], and other "evolutionary" algorithms also make use of randomization. In order to understand what is distinctive about genetic algorithms it is best firstly to concentrate upon the selection and mating processes occurring within them. In search or optimization problems these processes are usually applied directly to candidate solutions to a problem. In machine learning systems, for example in classifier systems, these processes are applied to a population of rules for solving a problem. In this chapter I shall be primarily concerned with the use of genetic algorithms for search and optimization. A search or optimization problem is suitable for the application of genetic algorithms if:

(a) It is one for which the search space is large.

(b) It is one for which candidate solutions can be uniformly represented.

(c) These solutions can be placed in a transitive order, as better or worse attempts to solve the problem.

(d) New candidate solutions can be generated by combining together elements of previous candidate solutions.

In a standard genetic algorithm the candidate solutions are represented by bit strings, but other forms of representation are used for various subject domains. The measure which is used for ranking is customarily called "fitness", stressing the analogy with natural selection. A genetic algorithm works by:

(1) Initializing with a population of candidate solutions - in many implementations this initialization is done by purely random selections from the search space, but initialization can be done using a variety of simple heuristics. N.B. It is important that initialization produces a diverse population of solutions.

(2) Ranking the population of candidate solutions according to the criterion of "fitness" being used.

(3) Generating a new population by

   (i) Performing selective mating upon the population of candidate solutions: the fitter a candidate is, the greater the likelihood of mating and therefore the higher the probability of offspring.

   (ii) Performing other operations on candidate solutions, e.g. random mutation, inversion of substrings: again, the fitter the candidate the greater the number of offspring.

(4) Repeating (2) and (3) until a sufficiently good candidate has been produced or until repetition has reached a set limit.

By this process a genetic algorithm is meant to isolate the advantageous components of candidate solutions and, by recombining them to form novel solutions, it is meant to identify improved solutions. A key feature of the strategy is the idea that there may well be a component which is highly advantageous in candidate solutions which are, for example, 20th from the best known candidate solution to a problem. Many other search and optimization algorithms fail to make any use of the advantageous features of such also rans. Genetic algorithms have been used successfully in a wide range of search and optimization problems, and combined with other algorithms they have featured in successful machine learning systems, for example a prototype of a system for controlling a gas pipeline. Moreover a formal theory of their operation has been developed by J.H.Holland [Holland, 1974], who is indeed the main figure in the development of genetic algorithms. [Goldberg, 1989] is an extensive introduction to genetic algorithms.

## 3.2 An Example of a Hybrid GA for Discovering (Logical) Models

I shall now present an example of a genetic algorithm. It is not a pure genetic algorithm. It is a hybrid genetic algorithm: it combines together features of a genetic algorithm with features related to logic. This involves the use of non-standard mating (crossover) operators. The algorithm is my own design and I call it METAGEN.1. This stands for METAlogical GENetic algorithm version 1. Later in the chapter I shall explain how this hybrid algorithm may be related to scientific method. For the moment let us just consider it as a simple example of a hybrid genetic algorithm. To understand it we need to understand the problem on which it runs.

This problem is the SAT (Satisfiability) problem. To a pure logician unconcerned with computational issues this problem may seem oversimple. Certainly it lacks the logical structures which are needed to express scientific theories in all their rich detail. Nevertheless it is a problem which is recognized in the theory of computational complexity as a highly complex problem. SAT is an NP-complete problem (see below for a definition of NP-complete). It is the problem of whether a set of clauses, in the propositional version of clausal logic, has a model or not. A clause is just a disjunction of literals, i.e. of positive or negatively signed atomic propositions. Thus "a or not b or c" is a clause. Let us briefly review the reasons why SAT is a computationally complex problem. Consider a simple exhaustive search algorithm for solving SAT for a set of clauses where the number of atomic propositions appearing in the clause set is $n$. Exhaustive search proceeds by systematically trying every possible truth assignment for the $n$ atomic propositions, and checking whether each such truth assignment provides a model for the set of clauses. Since there are two truth values, and $n$ atomic propositions, the number of truth assignments to be considered is 2 to the power $n$, where $n$ is the number of atomic propositions. Thus, as $n$ increases in size, the time taken for exhaustive search on a sequential machine will grow exponentially. Thus, considered as a general algorithm for solving SAT, exhaustive search of the space of truth assignments is a bad algorithm, because the time it takes crosses the polynomial time bound; that is to say, for any $k$, there will be some $n$ for which the exponential time that the algorithm takes will be greater than $n$ to the power $k$.

The importance of this can be seen by considering the following table, which assumes that one loop of a program takes a microsecond to execute:

| Time Taken to Execute L Loops | | |
|---|---|---|
| | $n = 10$ | $n = 60$ |
| $L = n$ | .00001 seconds | .00006 seconds |
| $L = n^5$ | .1 seconds | 13.0 minutes |
| $L = 2^n$ | .001 seconds | 366 centuries |

Thus we can see that it is too costly to attempt to solve SAT problems by exhaustive search for truth assignments, even for problems involving just 60 atomic propositions. More importantly than this we know that SAT is NP-complete. Therefore we have good reason to believe that there is no algorithm that provides a guaranteed polynomial time solution to all instances of the SAT problem.

To understand this consider the proof strategy used for the proof that SAT is NP-complete ([Cook, 1971]: see [Garey and Johnson, 1979] for an account). This proof concerns a Non-Deterministic Turing machine, that is one that has a "guessing" module which can randomly write any string from a specified alphabet, and which otherwise behaves like an ordinary Turing machine. We say that a decision problem is soluble with a "yes" answer in Non-Deterministic Polynomial Time provided that a Non-deterministic Turing Machine has a non-zero chance of guessing the solution to that problem, and checking that it is a solution, within polynomial time. Thus, for example, if there is just one model to be found for a SAT instance of size 60, then a Non-Deterministic Turing Machine programmed to make just one guess and then check it has less than a one in a million million million chance of guessing the solution to that problem, but it still does have more than a zero chance so we say that the SAT problem is soluble in Non-Deterministic Polynomial Time by this machine.

A problem is NP-complete if and only if (1) it is soluble for the case of a "yes" answer in Non-Deterministic Polynomial Time and (2) any problem that is soluble in Non-Deterministic Polynomial Time is reducible to it by a (deterministic) polynomial time algorithm. We have already seen that condition (1) is met for SAT. The proof of (2) for SAT [Cook, 1971] is fairly long-winded but the idea behind it is simple enough. The idea is that the conditions for successful running and halting of any Non-Deterministic Turing machine on any problem can be stated in a set of SAT clauses and it can be proved that the number of SAT clauses needed is a polynomial function of problem size. Since Cook's proof that SAT is NP-complete, many other problems have been proven to be NP-complete, typically by transformation from SAT.

The importance of NP-completeness results depends upon the conjecture that P $\neq$ NP, i.e. that there are some problems soluble in Non-Deterministic Polynomial Time which are not soluble in (deterministic) Polynomial Time. If the conjecture holds true then it follows that no NP-complete problem is soluble in P. This conjecture has not been proven to be true, but it seems highly plausible. To see its plausibility consider the relationship between a Non-Deterministic Turing machine and a parallel machine which is allowed to branch at each step using as many processors as its branches require. Anything that this parallel machine can solve in polynomial time can be solved in Non-Deterministic Polynomial time. Thus we might run our parallel machine on the SAT problem. Suppose it required a processor for each interpretation it generated for the SAT problem, then the number of processors would rise exponentially with the size of the problem. Consider the SAT problem with 60 atomic propositions. Then it would require more than a million million million processors, which is over ten million times more processors than there are estimated to be neurones in a typical human brain [Rummelhart, 1987, p.131]. It is implausible that anything that can be solved in polynomial time by a machine which can use processors at an exponentially growing rate can be solved in (deterministic) polynomial time by one processor operating alone. Yet this is what one has to suppose if one supposes that P=NP. If P $\neq$ NP then in practice we need to look for heuristic solutions for SAT and for other NP-complete problems.

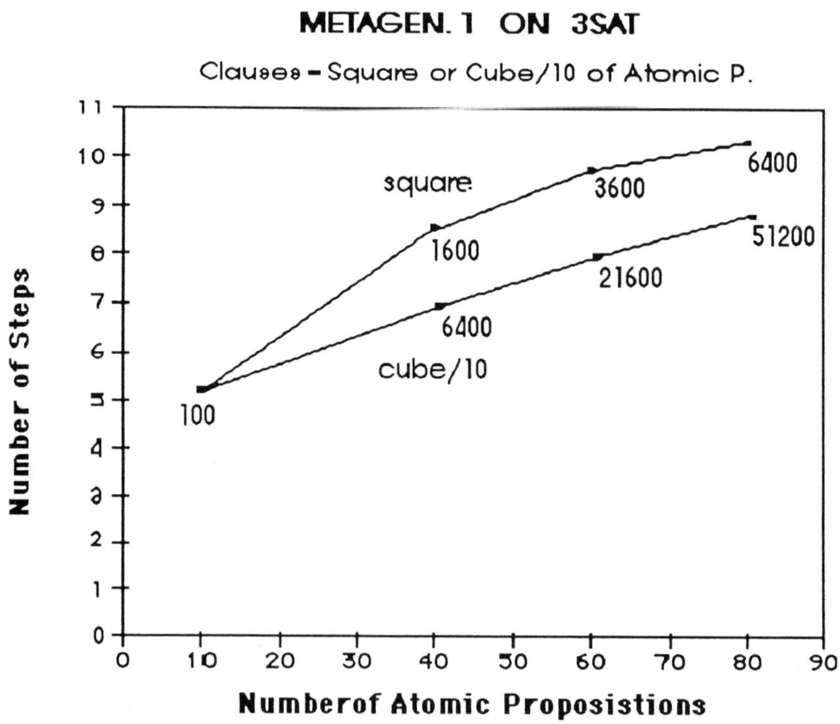


Graph 1

| TABLE 1: METAGEN.1 ON 3SAT | | | | |
|---|---|---|---|---|
| Number of Clauses is Square of Atomic P | | | | |
| Number of Atomic Propositions | 10 | 40 | 60 | 80 |
| Number of Clauses | 100 | 1600 | 3600 | 6400 |
| Number of Runs | 100 | 100 | 100 | 100 |
| Number of Successful Runs | 99 | 100 | 100 | 100 |
| Average Number of Steps | 5.20 | 8.55 | 9.75 | 10.31 |
| Minimum | 1 | 5 | 7 | 8 |
| Maximum | 45 | 12 | 14 | 17 |
| Number of Clauses is Cube/10 of Atomic P | | | | |
| Number of Atomic Propositions | 10 | 40 | 60 | 80 |
| Number of Clauses | 100 | 6400 | 21600 | 51200 |
| Number of Runs | 100 | 100 | 100 | 100 |
| Number of Successful Runs | 99 | 100 | 100 | 100 |
| Average Number of Steps | 5.20 | 6.95 | 7.9 | 8.81 |
| Minimum | 1 | 5 | 6 | 7 |
| Maximum | 45 | 9 | 10 | 11 |

In Graph 1 and Table 1 we can see how METAGEN.1 fares on instances of the 3SAT problem. It is run on the 3SAT problem, where the size of each clause is restricted to 3 literals. But since the question of whether any arbitrary set of SAT clauses has a model can be reduced to the question of whether a set of 3 SAT clauses has a model, this restriction does not matter. (Consider the clause "a or b or c or d": it has a model if and only if the two clauses "e or a or b" and "not e or c or d" have a model). Instances of the 3SAT problem are generated by randomly constructing an interpretation (set of truth assignments to atomic propositions) and then randomly constructing a list of clauses with the constraint that each clause is to be true according to that string of truth assignments. In order to save time in construction duplicate clauses are not removed from the list, so the number of clauses in the list is somewhat larger than the number of distinct clauses in the clause set for each problem. However, if the number of atomic propositions is $n$, then the chance of a duplicate occurring for each clause is approximately

$3/(2 \times n)$ when the number of clauses in the clause set is the square of $n$, and 3/20 when the number of clauses is one tenth of the cube of $n$.

Table 1 and Graph 1 record the results of a series of experimental runs on METAGEN.1 which were undertaken for me by an M.Phil. student A. Reel, who has also worked with me on the detailed programming of the algorithm. We intend to publish a fuller account of these results in a joint paper at a later date. The graph shown here records the results of 800 runs of the algorithm – 100 on each of the 8 data points recorded on the graph. Rather than recording the time taken by the algorithm we have recorded the number of steps taken by METAGEN.1 to reach a solution (the meaning of "step" is given on page 42). This number of steps is low considering the problem being handled. Thus METAGEN on average solves even the 80 atomic proposition problems with 6400 clauses or 51,200 clauses in approximately 10 steps. Out of the 800 runs recorded on the graph there was just one run in which METAGEN.1 failed to discover a model. It converged too quickly on a false optimum. Reruns of METAGEN.1 on the precise set of clauses on which it had failed, but with a different seeding of the random number generator for the genetic algorithm showed that it could solve this problem, indeed it solved it 98 times out of a 100 in our further tests. However, it must be confessed that we also tried METAGEN.1 on a problem which had caused difficulties for an earlier (somewhat bug-ridden) version of METAGEN, and we found that METAGEN could only solve that problem in about half its runs (when it did solve the problem it solved it quickly). Interestingly, both these problems were of the 10 atomic propositions with 100 clauses size.

METAGEN is capable of a high degree of parallelism and therefore the number of steps it takes is a reasonable measure of the time METAGEN needs to identify models. Thus it might be a reasonable account of the time taken by an implementation of METAGEN in a neural network. But would a parallel METAGEN require an enormous explosion in the number of parallel processes as the number of atomic propositions grew in size? So long as the number of clauses did not increase at a more than linear rate the number of processes required by METAGEN would grow only linearly. If the number of clauses grew such that it was a constant proportion of the number of possible distinct clauses then, for 3SAT, it would only increase with the cube of the number of atomic propositions. The processes that are being considered here may each be modelled as a process acting on one synapse, and thus, in a neural net, one neurone might handle many such processes. Thus a small proportion of the human brain would be needed to handle problems of the size under consideration on the table.

How does METAGEN.1 achieve its success? The algorithm used is essentially very simple. A database of 20 candidate solutions (strings of truth assignments) is kept in an array (the number of database candidates is held constant at 20 for any size of problem). Associated with each member of the database there is a pair of other array segments that represent candidate solutions that are in competition with it. At initialization the database and the associated set of pairs are all randomly initialized. At each further step for each database candidate a new competing pair of candidate solutions is created (previous competitors being overwritten). One of these new solutions is created by a special mutation operator operating on the database candidate with which it is paired. The other is created by a special crossover operator which operates on a randomly selected pair of database candidates. The fitness of a database candidate and its competitors is judged at each step, and the fittest (or equally fittest) candidate for each such triple is the new database candidate for the next step. Fitness is judged by the number of clauses in the clause list that a candidate makes true.

At the heart of the algorithm are the special mutation and crossover operators. These make special use of information derived when the fitness of a candidate is assessed. Consider the $n$th truth assignment (i.e. the truth value for atomic proposition $n$) in a particular candidate solution. Consider the mutation operator operating on the value for proposition $n$. It uses information about the number of clauses ($Fn$) that are actually made false by that truth assignment in the context of its database candidate solution, and also about the number of clauses ($Cn$) in which proposition ($n$) is to be found. A random number ($Rn$) from 0 to $Cn$ is selected and if $Rn \leq Fn$, then the value of proposition $n$ is changed. Thus the higher the proportion of its clauses that the truth value of proposition $n$ makes false the greater the chance of its truth value being changed. At the same time there is a slight chance of change even if $Fn$ is 0, i.e. if the assignment makes no clauses false. At each step the mutation operator for a database candidate performs this operation on each truth value in its truth string. Now consider the mating or crossover operator. It works on two database candidates. Let $Ta.n$, $Tb.n$ be the truth values for the $n$th proposition in candidate solutions $a$ and $b$ and let $Tc.n$ be the corresponding truth value for the new candidate. If $Ta.n = Tb.n$, then $Tc.n$ is assigned the same value. Otherwise, if $Ta.n$ makes $Fa.n$ clauses false in the context of its database candidate and $Tb.n$ makes $Fb.n$ clauses false in the context of its database candidate, then a random number is chosen from 0 to $(Fa.n + Fb.n + 2)$ and, if this random number is less than $Fa.n + 1$, then $Tb.n$ is selected. Thus the greater $Fa.n$ is with respect to $Fb.n$, the more likely it is that $Tb.n$ will be picked. Again there is a slight chance of $Ta.n$ being changed even though it makes no clauses false.

| TABLE 2: METAGEN.1 ON 3SAT | | | | |
|---|---|---|---|---|
| Number of Clauses is Square of Atomic P | | | | |
| Number of Atomic Propositions | 10 | 40 | 60 | 80 |
| Number of Clauses | 100 | 1600 | 3600 | 6400 |
| Number of Runs | 100 | 100 | 100 | 100 |
| Number of Successful Runs | 99 | 100 | 100 | 100 |
| Average Successful Crossovers<br>10 Crossover Attempts | 3.287 | 5.768 | 6.334 | 6.694 |
| Average Number of Mutations<br>10 Crossover Attempts | 1.990 | 1.472 | 1.310 | 1.215 |
| Total Successes from Genetic Operator<br>10 Opportunities | 5.277 | 7.240 | 7.645 | 7.909 |

Note that each operator on a candidate is a set of independent processes, one process for each truth value. This makes sense because from the number of clauses a truth value makes false, we know the number its opposite will make true, since clauses are disjunctions. However, with these independent changes more truth values may change than are necessary to cause false clauses to be true and too great a cost may be payed in true clauses becoming false. The probabilistic nature of the algorithm compensates for this. Moreover the independent local decision making in the operators provides a high degree of potential parallelism. The probabilistic nature of the operators facilitates this, because it allows decisions to be made locally without costly communication.

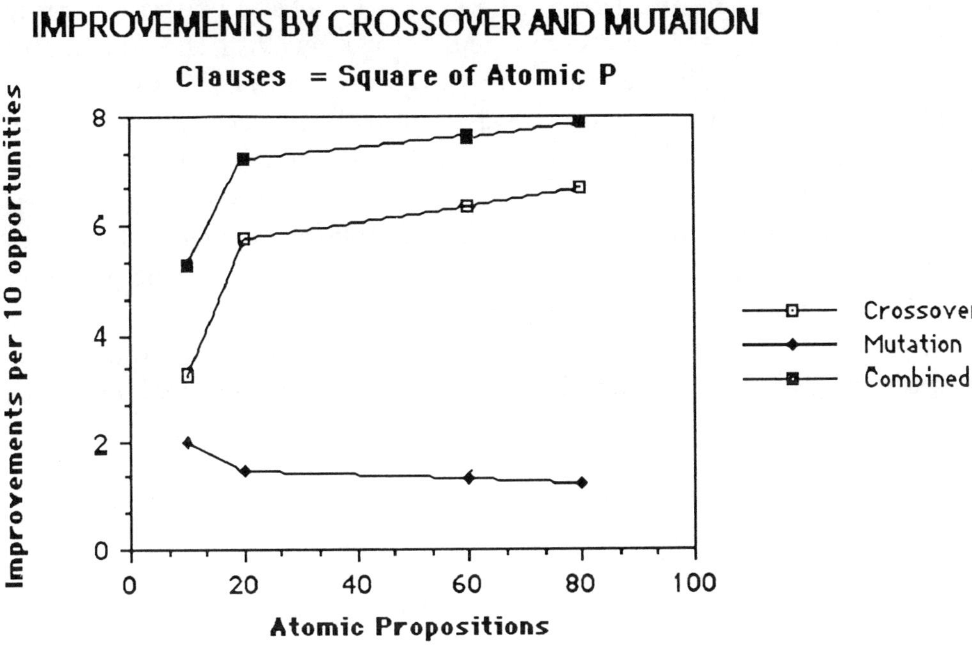


**Graph 2**

These two operators prove very effective in practice, as can be seen from Graph 2 and Table 2. On the largest problem size considered the crossover operator effects improvements in database candidates on average 6.7 times for every 10 attempts it makes. Another way of looking at things is that on the smallest problem size METAGEN tends to look at approximately 260 interpretations or a quarter of the possible interpretations before arriving at an answer, but on the largest problem size it only looks at approximately 460, which is a minute fraction of the interpretations available. The mutation operator is not as effective as the crossover operator, although it comes close to its performance on small scale runs. The mutation operator can be expected to be less successful as the algorithm moves towards a conclusion because, when a truth assignment to a proposition makes only a very small proportion of clauses to which it is relevant false, there is only a very small chance of change through mutation.

How does METAGEN relate to scientific method? Surely it lacks the rich data structures that are necessary if one is to express scientific theories? Surely the predicate calculus together with the addition of some mathematics is necessary for expressing scientific theories? It is of course true that we cannot state general scientific theories using only the propositional calculus. Nevertheless METAGEN illustrates how logic and genetic algorithms may be combined. Perhaps more sophisticated combinations will prove to be possible. Moreover it illustrates the use of a heuristic non-deterministic algorithm for a logic problem. Philosophy of science needs to consider the computational complexity of problems in scientific method. Heuristic and non-deterministic processes may be necessary in scientists' practical judgement making. Non-deterministic cognitive processes would allow scientists to diverge in judgements without diverging in fundamental method. Another point is that the overview a scientist has of a field may be expressible in SAT clauses, e.g.:

> "Either Theory T is false or Auxiliary Hypothesis A is false or Hypothesis H is false"
>
> "Either Hypothesis H is true or Hypothesis B is false or Hypothesis C is false"

and so on. Clearly, with enough such clauses, the SAT problem becomes serious. The problem might grow serious for an individual let alone for a group of scientists. METAGEN can provide the core of a cognitive science theory of how scientists cope with this problem.

It may also be able to contribute to an account of what it is rational to do when confronted with the problem of judging which is to be selected from a subset of a set of plausible hypotheses. Rescher [1976, p.53] proposed that in reasoning with plausibility we ought to determine whether a set of complex propositions under consideration is itself consistent, and, if it is not consistent then we ought to determine its set of maximally consistent subsets. We are then supposed to choose between these sets on the principle of choosing the set which excludes only propositions of relatively low plausibility. If there is a dead heat in this respect then we are supposed to accept only those propositions which are consequent upon the equally plausible subsets. A problem with Rescher's proposal is that, as we have seen, the SAT problem is computationally complex. Rescher's proposal can only be applied in practice if a heuristic algorithm is used for searching for maximally consistent subsets. SAT seems to be a contender for this role.

A question we need to consider is the computational complexity of theory selection in science. The SAT problem shows us that even if in particular cases we can reduce problems of consistency, discovery of logical modelling etc. to problems in the propositional calculus, we are still left with problems of great computational complexity. No doubt there are other problems which are of equal or worse complexity, both at the general level (e.g. predicate calculus) and in particular scientific disciplines. Genetic algorithms may well be applicable to many such problems. But there is a more general question that can be posed: can genetic algorithms provide such a considerable contribution to epistemology that it makes sense to talk of an evolutionary epistemology? In a full discussion of this issue it would be necessary to consider in detail the variety of different kinds of genetic algorithms that are available, including machine learning algorithms, but since there is not time for that in this chapter I shall review some general philosophical arguments that have been put forward by Thagard against evolutionary epistemology [Thagard, 1988].

## 3.3 GAs as a Basis for Evolutionary Epistemology

The question that we need to address is: Could Genetic Algorithms provide a basis for an account according to which human knowledge develops, *indeed develops rationally*, by a process that is in some deep sense akin to a process or processes that occur in biological evolution? In his book *Computational Philosophy of Science* Thagard [Thagard, 1988, Ch.6] argues that evolutionary epistemology cannot usefully contribute to philosophy of science. Strangely, in his attack upon evolutionary epistemology, Thagard ignores genetic algorithms entirely and concentrates his attack upon Campbell's arguments for evolutionary epistemology. Yet Thagard is not ignorant of genetic algorithms, indeed he is co-author with J.H.Holland et al. of a book on Induction in which they feature extensively [Holland et al. 1986].

Thagard recognizes three respects in which the development of scientific knowledge is very loosely analogous to biological evolution. They both can be said to develop by means of variation, selection, and transmission. But Thagard argues that when we examine each of these respects in detail, we find that scientific development is very different from biological evolution. (N.B. Some of the following descriptions of Thagard's arguments are my own summaries; some are direct quotations.)

## Thagard 1: Variation

**(T1.i)** ([Thagard, 1988, p.102-105]) In biological evolution we have random mutation, which is what Campbell [Campbell, 1974, 1977] concentrates upon, but the development of scientific knowledge is a rational development of prior knowledge, and, although conceptual combination is superficially like crossover of genes in sexual reproduction, it is actually much more constrained than this. Thus Thagard argues (op.cit. p.104), that if a modern detective formed hypotheses by a process akin to crossover in sexual reproduction, then he would be as likely to form the hypothesis that Attilla the Hun was the murderer as to form a useful hypothesis about a contemporary individual.

Hybrids like METAGEN which combine genetic operators with logical (etc.) considerations may offer the constraints which Thagard argues that evolution lacks. Even from the perspective of standard genetic algorithms this criticism seems misplaced. The application of a genetic search algorithm in the work of the police would be to a much more constrained search space than that envisaged by Thagard. We can envisage the police working with a genetic algorithm that is restricted to the individuals known to have been in a location, plus, perhaps, a small set of unknown strangers; and which is attempting to identify the best supported case against a suspect which is consistent with what is known about the killing and with what is known about the suspect. Perhaps a case has been built up against one suspect which is highly plausible and it founders because that individual has a cast iron alibi. The individual is a burglar who is known to enter "burglar proof" buildings in a certain highly efficient manner, and this seems a plausible way for the murderer to have entered the building. Now the police realize that elements of their case against the burglar can be combined together with elements of a case being built against another suspect. If this other suspect (not a known burglar) is credited with knowledge of the burglar's technique, then a highly plausible case can be built against him. It is just this kind of recombination that a genetic algorithm might help us to effect.

Other arguments that Thagard puts against the strength of the analogy between variation in the development of scientific knowledge and variation in biological evolution are:

**(T1.ii)** (op.cit. p.106) that 'in the history of science variation and selection are "coupled": the factors responsible for selection are related to those responsible for the original generation of variants' whereas in biological evolution 'species variation and selection are "uncoupled": the factors that produce genetic change are unrelated to the environmental struggle for survival, except in special cases where the environmental threat is unusually mutagenic',

and

**(T1.iii)** (op.cit. p106-7) that in 'scientific development .... the rate of theoretical variation seems to be partly dependent on the degree of threat to existing theories', whereas 'biological variation is not similarly sensitive to degree of environmental pressure on organisms'.

Whilst these criticisms may be relevant to versions of evolutionary epistemology that have been put forward in philosophy of science, they do not serve as cogent arguments against an evolutionary epistemology based upon genetic algorithms. It is noteworthy that in argument (T1.ii) Thagard concentrates upon "species variation". This is not the analogy upon which standard genetic algorithms are based. They are concerned with variation occurring within the population of a species allowing different matings to produce variation amongst individuals of that species. Here variation can occur because of the creation of polygenetic complexes. The fitter an individual is the more likely it is to have offspring. Here the likelihood of a novel kind of individual being produced is dependent upon the fitness of the individuals whose mating is capable of generating that novel variation. Thus there is a systematic coupling between variation and fitness in genetic algorithms. It is not a coupling which guarantees that every new variation will have improved fitness, but it is a coupling that offers a fruitful way of searching for improved variations.

A similar point holds against Thagard's argument (T1.iii). If we consider the rate of variation produced through matings in a population, then this may slow down in a situation where the population is well adapted to its environment and stable, because of a tendency for one or a few genotypes to dominate in such a population, giving little possibility of variation through recombination. On the other hand, when the population has not settled like this there will be a greater tendency for novel variations to be developed because the population will not yet have stabilized on a few genotypes. Clearly there is a danger here of a run of a genetic algorithm becoming unproductive because of too early a domination by a small number of genotypes. This can happen in applications in search and optimization, where fitness can be measured definitively by a simple algorithm. There is even greater likelihood of this happening with scientific theories where a theory that once had a good fit with the data ceases to have a good fit because of new evidence. Happily protection against this problem can be built into genetic algorithms. One way of doing this is "niching". Here we make special provision in our selection process for a diversity of genotypes to be preserved. We will ensure that there are individuals in the mating population to fit each niche. This is consistent with allowing the fittest individuals to be mated much more often than individuals who are allowed to mate merely to preserve a niche. In the case of search and optimization niching can provide some degree of variation even when a few genotypes vastly outclass their competitors in terms of fitness. In the case of scientific theories, where there is a sudden perceived drop in the relative "fitness" of the dominant theory, because of new data, what will happen is that it will be reproduced far less often in the population and therefore variations arising from novel combinations amongst hitherto lowly niche-fillers will have a much greater chance of proliferation. In other words we will get the result that Thagard says we could not get, namely that there will be an increase in variation amongst our population when the genotypes dominating the population are under threat. Moreover, genetic algorithms gain this result by developing the biological analogy, not by moving away from it.

## Thagard 2: Selection

**(T2.i)** (op.cit. p.107-9) General Standards: In science we have general standards that enable us to discriminate between successful theories and unsuccessful ones, whereas in biological evolution what makes a thing fit varies from environment to environment. Thus it makes sense to talk of general progress in science, but this concept has no valid application in biological evolution. Therefore the analogy between selection in science and selection in biological evolution is very slight.

This point about general standards does tell against an analogy between science and the whole of biological evolution. But the analogy that is used in genetic algorithms is not an analogy between the process of the algorithm and the whole process of biological evolution. It is an analogy between a problem solving process, which presupposes a sharply defined criterion for whether the problem has been solved, and the process whereby the population of a species becomes adapted to a specific environment. Thagard's argument does not bear on this analogy. As before his argument may be a cogent criticism of some accounts of evolutionary epistemology in the philosophical literature, but it is not a cogent criticism of an evolutionary epistemology based upon genetic algorithm theory.

## Thagard 3: Transmission

**(T3.i)** (op.cit. p.109) "A beneficial gene is replicated in specific members of a population, but a successful theory is immediately distributed to most members of a scientific community. Preservation is by publication and pedagogy, not by any process resembling inheritance. Dissemination of successful theories is much more rapid than dissemination of beneficial genes. ... Thus at the level of transmission of units of variation .... the growth of knowledge is different from the evolution of species".

This is true, but in a standard genetic algorithm mating is done selectively: the fitter the individual the more matings it is likely to have. (N.B. At present METAGEN.1 mates all individuals that are selected for its database equally, and therefore it dispenses with this standard feature.) Depending on the function that relates fitness and matings, the spread of a beneficial gene can be much more rapid than it would be in most biological populations. Indeed this can give rise to a problem of too early a concentration of one genotype. As we have already seen too rapid a spread can lead to lack of variation in further generations, an effect which we can avoid by the use of niching in our genetic algorithm. In scientific communities too we can have a bandwagon effect in which an apparently good new theory becomes the basis of the research of a majority of scientists, but meanwhile there are theories being ignored which, if only they were further developed, would lead to much better theories. Indeed reflection on this point reveals an equivocation in Thagard's discussion of theory transmission. It is indeed often true that knowledge of apparently superior new theories is transmitted quickly in a scientific community; and no doubt this ought to happen in scientific communities. But it does not necessarily follow that research commitment to that theory spreads as rapidly or as widely, and it may well be that limits ought to be set upon the extent to which research commitment to it does spread.

**(T3.ii)** (op.cit. p.109) 'Genes serve as templates for new genes, which produce organisms. In contrast, theories and their components do not produce new theories. To use Hull's simile, a theory is like a parasite inhabiting those who hold it and capable of spreading to others'.

But theories are often constructed by combining elements of earlier theories. Thus they do act as templates and are different from parasites.

**(T3.iii)** (op.cit. p.110) 'Dawkins (1976) postulates "memes" as the conceptual replicating entities analogous to genes. But this postulation is gratuitous since we already have notions that describe the entities that develop in scientific and cultural change. These entities include rules, theories, laws, data, concepts and so on'.

Perhaps, but in genetic algorithms forms of data representation are used which do not presuppose that there are completely novel fundamental replicating entities. The application of genetic algorithms to the development of scientific theories need not involve fundamental replicating entities, which are quite different from syntactic representations of rules, data etc. What the application of genetic algorithms does require is some standard form of representation which can be used to represent any of the variety of theories that might usefully be applied to a subject domain. This may be feasible for narrowly defined subject domains, but is it feasible when we do not have a narrowly defined subject domain? This does present a problem for a general evolutionary epistemology. But equally it presents a problem for a general computational philosophy of science. J.H.Holland designed what he described as a "broadcast language" and, as he says, [Holland 1974, p.158]

> "it is possible to devise languages - the broadcast language is an example - which use strings to rigorously define all effectively specifiable representations, models, operators, etc. Since the objects of the language are represented as strings, they can be made grist for the mill provided by genetic plans."

Thus there is not a problem with the formal expressive power of the representations available for genetic algorithms. There is a problem about alleged incommensurability of theories etc., but this is a problem that faces any general computational philosophy of science; indeed it faces any general methodology of science.

**(T3.iv)** (op.cit. p.110) We may think in terms of a social evolutionary epistemology "with scientists as individuals in whom ideas arise, with transmission of ideas being like the spread of a biological trait in accord with the principles of population genetics", but sociology of science reveals that transmission of ideas is influenced by "scientific elites, communication networks, and other social organizations that have properties not found in populations of species other than humans (Crane 1972)." Parallel computation is a better theoretical metaphor for this than population genetics.

Undoubtedly there are social phenomena in human scientific communities which do not have correlates in the phenomena of the population genetics of simple animals, and surely some of these social phenomena are conducive to the development of science, although I expect that there are also many that are not. The question that we must address is: "Will genetic algorithms, based as they are on an analogy with population genetics, be useful in developing models of group rationality in scientific communities?". If we try to decide this question by deciding that parallel computation is "a better theoretical metaphor" than "population genetics" then we are in a muddle, because genetic algorithms are themselves intrinsically parallel and provide a form of concurrent computation that has been implemented on a variety of parallel computer architectures (G. Robertson in [Davis, L., 1987], papers in [Grefenstette, J. J., 1987, 123-183]). Therefore these metaphors are far from being exclusive of each other. It is implausible that a pure genetic algorithm could by itself provide all that is needed in a parallel computational model of group rationality in science. Therefore the question that we need to address is the question of whether parallel computational models that incorporate genetic algorithms, or that incorporate features of genetic algorithms, have advantages over those that do not. This is a question that can only be settled by a combination of analysis of different computational models and of experiment with them.

**Acknowledgments**
The idea of METAGEN came to me during my year as a Senior Visiting Fellow in Computer Science (Declarative systems Project) at the Uni-

versity of East Anglia in 1987-8. Professor M.R.Sleep encouraged me to work on genetic algorithms. I also learned more about the field whilst I was exploring neural networks and genetic algorithms for the TSP whilst on a visiting Fellowship at British Telecom Research Laboratories in 1988. As I have stated in the paper Aidan Reel has worked with me on the implementation and testing of METAGEN.1. We intend to publish a joint paper about it.

## References

Campbell, D.T., [1974]: "Evolutionary Epistomology." in Schilpp, P. (ed.), *The Philosophy of Karl Popper*, La Salle, 1h: Open Court, 413-463.

Campbell, D.T., [1977]: Comment on "The Natural Selection Model of Conceptual Evolution." in *Philosophy of Science* 44, 502-7.

Cook, S.A., [1971], "The Complexity of Theorem Proving Procedures," in *Proceedings of Third Annual ACM Symposium on the Theory of Computing* 151-8.

Crane, D., [1972]: *Invisible Colleges: Diffusion of Knowledge in Scientific Communities.* University of Chicago Press, Chicago.

Davis, L., [1987]: *Genetic Algorithms and Simulated Annealing*, Pitman. London.

Garey, M.R., and Johnson, D.S., [1979]: *Computers and Intractability: A Guide to the Theory of NP-Completeness*, W.H.Freeman. New York.

Grefenstette, J.J., [1987]: *Genetic Algorithms and their Applications*, Lawrence Erlbaum. New Jersey.

Goldberg, D.E., [1989]: *Genetic Algorithms*, Addison Wesley. Reading, Ma.

Holland, J.H., [1975]: *Adaptation in Natural and Artificial Systems*, Ann Arbor: The University of Michigan Press

Holland, J.H., Holyhoake, K.J., Nisbett, R.E., and Thagard, P.R., [1986]: *Induction*, MIT Press. Cambridge, Ma.

Rummelhart, D.E., and McClelland, D.L., [1987]: *Parallel Distributed Processing*, Vol.1 MIT Press. Cambridge, Ma.

Thagard, P.R., [1988]: *Computational Philosophy of Science*, MIT Press. Cambridge, Ma.

# Chapter 4

# Connectionism is Better for Engineers than for Scientists*

Derek Partridge
*Department of Computer Science, University of Exeter, Exeter EX4 4PT, UK*

There is a big difference between having a program that "solves" some problem, and having significant theoretical insights into the nature of the problem.

## 4.1 Preamble

The general question was "Does Connectionism Represent a New Paradigm for AI?" Several answers spring to mind: "yes", "no", and "who cares?" It is an answer that falls somewhere between the last two answers which I shall present. In short, I am very doubtful that connectionism is a useful, scientific paradigm for AI. The connectionist paradigm (CP) is not a paradigm that will significantly further the science of AI (if there is one). Perhaps the tail end of this assertion is

*This chapter is based on sections in several chapters of *A New Guide to AI* to be published by Ablex Publishing Corporation later this year.

better split into: further our understanding of human intelligence at the level of intelligent behaviour (i.e., cognitive science) or at the level of brain structure (i.e., neuro-biological science). The cognitive science level is more the traditional stomping ground of scientific AI, so I'll tend to concentrate on it.

The essence of my argument is that theories, models, and experiments must be effectively communicable within the local scientific community at least – otherwise we just don't have a science. At the root of this communication requirement is comprehension of complex theories and their relationship to even more complex models. The combination of connectionism and computers exacerbates this communication problem to the point where it is difficult to believe that such a scientific community could indeed exist. The necessary accumulation of abstract theoretical insights, at the level of cognitive phenomena, independent of any particular program but testable in any number of possible programs, would appear to be especially unlikely when the models are CP-based.

## 4.2 Overview

First, I'll outline the major methodological framework within which models and theories are developed in AI: models are programs, and executing these programs is experimenting with the model which leads to development of the underlying theory. A quick review of the points for and against the CP is given next. It is within this part of the chapter that the arguments *vis-a-vis* connectionism as a paradigm for both cognitive science and neuro-biology arise and are discussed. Machine learning is then considered (and rejected) as a means for CP-based systems to sidestep many of the problems raised. Next, I'll widen the focus of the chapter to point out why connectionism illuminates the general problem of "Is AI a science?", and, if so, "What sort of science is it?" And finally, there are some positive suggestions to help balance the thoroughly negative bias of all that has gone before.

## 4.3 AI System Development

Exploratory programming is one of the phenomena to emerge from work in AI. This term refers to the use of a computer together with a programmed model to explore and develop the model (or some underlying theory embodied in the model) such that it becomes a more adequate model (and thus a more adequate theory). This is the scientific slant on exploratory programming. For the engineer, exploratory programming is a means by which the adequacy of a potential piece of applications software can be examined and developed in order to finally yield a useable product. A significant point of contrast between these two views of what is obstensibly much the same procedure is that for the scientist it is the insights into the general phenomenon that are of primary interest, whereas for the engineer it is the behaviour of the specific program at hand.

But in both cases progress is made through a process of incremental development of a specific program. This process can be characterised as the run-understand-debug-edit cycle, or more memorably, the RUDE cycle. Within this RUDE cycle it's the "U" that's particularly worrying, and leads on to problems with the "D" and the "E". If we can't develop a satisfactory understanding of the behaviour of our program, then there's not much chance of modifying it in a controlled and disciplined way. In this situation the RUDE cycle will only exist in its worst manifestation – more or less random code hacking – and the likelihood of steady development and thus convergence on a state of adequacy will be somewhat less than random chance. So lack of understandability in our programmed artefacts spells trouble for both the scientist and the engineer.

An obvious, and not uncommon, rejoinder to this argument requires us to admit the unsatisfactory nature of RUDE and so eliminate this understandability problem by claiming that it stems from a misguided notion of computer-program development. Building AI systems should be more like the design of an algorithm to implement a well-specified problem as CS centrists advocate, not speculative modification of an inadequate program in the hope of chancing on an ill-defined goal. Specify the problem before you start programming; design an algorithm to implement this specification; and finally verify (or at least exhaustively test) that the programmed algorithm does indeed implement the specification. "That's the only sensible way to develop computer systems," the specify-and-verify enthusiasts (the SAVs) will tell you.

This argument will take us far afield if we're not careful (and I'd like to go there, but there is no space to do it in this chapter – the interested reader is invited to browse [Partridge, 1986], for the full details, or [Partridge and Wilks, 1987], for a summary of most of the essential points). So I'll content myself with the argument that even if the SAVs are right (and they're not), it will not help the connectionist case. Why not?

If our aim is to merely (!) implement a specification correctly then whether we use connectionistic or thoroughly classical means to do so is almost irrelevant. We're no longer doing AI. Intelligence is, I believe, a family of approximate solutions to computationally intractable problems. The actual approximations instantiated in each of our heads will, almost certainly, vary from person to person. As to whether the individual approximations (or approximations of individuals) vary drastically or trivially, I haven't the slightest idea, but luckily this tricky issue is immaterial to the problem at hand and we can therefore drop it.

The point is that system building in AI necessarily involves the exploration of approximate solutions to otherwise intractable problems. This is totally different from the design and implementation of correct solutions to well-specified problems; it is also a lot harder.

Under this sort of rubric, the goal of cognitive science is to discover and explore the scope and limitations of the constituent approximations that altogether comprise intelligence. We must hope that there are significant core approximations which we all share, otherwise it is difficult to see the possibility of a science of cognition if we are each individually composed of radically different sets of approximations which in sum do more or less the same job – i.e., exhibit intelligence. This is presumably possible, and it is closely connected to what Fodor alludes to when he suggests that the mind may not be an appropriate phenomenon for scientific study (like Wednesdays in Bovey Tracey) – i.e., there are no interesting abstract regularities common to all manifestations of human intelligence.

I shall return to this larger issue when I have clarified the essential difficulty that an exploratory program development methodology implies, and disposed of the response that the power of CP-based learning algorithms eliminates much of the need for hand-crafted incremental system development.

## 4.4 A Need to Mess with the Innards

Predicting the future by examining the entrails of a slaughtered goat is a lost art, and that's a pity. Some insight into the nature of this intriguing process, which although not exactly analogous to procedures involved in incremental programming, could well be of use to the AI-system developer. For in AI, the necessary understanding of a programmed model boils down to the need to be able to evaluate the behaviour of the model and from it draw inferences about modifications to the structure that will both eliminate the undesired, observed behaviours and introduce desired, but unobserved, behaviours. In sum, we need to be able to predict structure from function. And we all know that this is an impossible task.

Without wishing to pretend that AI has found the philosopher's stone that solves this function-to-form problem, the very fact that the behaviour of all structural subcomponents of the system can also be explored does, I believe, substantially soften the impossibility claim. The system developer constructs an understanding of why the system is doing what it is observed to be doing by building up from an understanding of the constituent subsystems. And the conventional tenets of good programming practice provide considerable aid in this task of understanding the behaviour of a program.

The farrago of rules and guidelines that is collectively named structured programming is part of the response of conventional computer scientists to recognition of the software crisis in the late '60s (the term "software crisis" refers to the realization that the technology of the day, or lack of it, produced as a matter of course thoroughly unreliable and incomprehensible software systems). Management of complexity by modularisation within a hierarchical framework is a central pillar of the technology that computer science has developed in order to combat the software crisis.

Classical AI programming has been able to exploit much of this conventional wisdom in its programmed models. There are notable exceptions such as the brief fling with heterarchical architectures and a number of other manifestations of so-called "hairy" control structures (to use the technical term). Dependent upon the particular subdomain within AI, the classicists have felt more or less able to live with the constraints of the traditional technology for good programming practice.

In general, the classical theories of AI have been couched in terms that are not alien to the principles of good programming. As a result, models in classical AI are often constructable as programs that do little violence to the strictures of good programming. In addition, they are programs whose structure is a direct reflection of the structure of the theory under study. This means that such models provide considerable support for the difficult task of relating function to structure and to implications for the underlying theory.

In short, the systematic compositional structure of classical AI theories (a point that Fodor and Pylyshyn make much of, see below) can, by a competent programmer, be translated quite straightforwardly into a well-structured, and thus maximally comprehensible, program. The structure of the program reflects the structure of the underlying theory. Thus facilitating inspection of the penetralia of a programmed artifact in order to predict future behaviours as well as facilitate analysis of observed behaviour in terms of the innards of the system.

For CP-based models the story is rather different, but first we should briefly review this paradigm, if that is what it is.

## 4.5 Subsymbolic Connectionism: the Good News

Subsymbolic networks (i.e., networks within which single nodes have no semantic interpretation at the level of cognitive phenomena) have many attractive properties that seem to support the connectionistic claim that subsymbolic networks are not just low-level implementations of semantic networks (like machine code is a low-level but essentially equivalent alternative to a high-level programming language such as FORTRAN). Thus symbolic concepts such as 'apple' are not all-or-nothing entities in a subsymbolic network as they are in symbolic networks where the node labelled 'apple' is present (or active), or it is not. In a subsymbolic network this concept is represented by a general pattern of activity; one activation of the 'apple' pattern must be more or less similar to another, they need not be identical. Elimination of the necessity for all-or-nothing decisions accords well with our beliefs about intelligent reasoning, but it does raise the problem of constructing an effective general algorithm to compute sufficient similarity.

Apart from the possibilities for machine learning (discussed below) and the benefits of distributed representations, Smolensky [1987, 1988], in his exposition of the merits of the subsymbolic paradigm in AI, discusses the benefits of both parallelism and the use of 'soft constraints.' Parallelism has, in all subdomains of AI, been invoked as an ultimate enhancement, and furthermore, one that will rescue the known, but intolerably slow, techniques by providing some orders of magnitude increase in processing speed. Parallelism itself is another subfield of AI with much promise and many unsolved problems. Suffice it to say that connectionist systems are inherently parallel (activity values are broadcast in parallel from one set of network units to another), and thus they include the potential benefits of parallelism which appears to be an essential component of real-time intelligence.

[Waltz 1987] lists six reasons for the excitement that surrounds connectionist work:

1. It's a good candidate for a major new paradigm;

2. The similarity of connectionist networks to brain architecture suggests that they may have potential in cognitive science;

3. Connectionism holds promise for significant breakthroughs in machine learning (an AI subfield that constantly runs up against formidable obstacles in its classical manifestations);

4. Connectionist systems exhibit non-trivial fault tolerance, thereby circumventing the brittleness problem typical of classical systems (sometimes known as "the china syndrome" – see below);

5. Appropriate hardware is rapidly becoming available which is important for testing systems beyond the current very-small examples;

6. Connectionist systems also scale-up well: "modules can be interconnected rather easily. This is because messages passed between modules are activation levels, not symbolic messages."

This last claim is particularly interesting because it appears to be quite contrary to the views of [Hinton 1986], who claims that one of the problems is that scaling-up of the back propagation learning algorithm, for example, is not yet viable. The clash of these apparently contradictory claims is softened, if not avoided, when we realize that Waltz is referring to the construction of large symbolic connectionist systems by interfacing small-scale network modules, and Hinton is referring to large-scale subsymbolic networks.

## 4.6 When is an AI System like a Piece of Fine China?

There are at least two good answers to this curious question: always; and when it's a classical system. The brittleness of classical systems in AI is legendary (even if it is not universal). Occurences of the china syndrome plague most classical systems: they either get it dead right or fail miserably, there are no mediocre performances.

(The china syndrome is not to be confused with the China syndrome: the not uncommon situation in AI where we cannot reproduce the fancy results that have been published because the only person who can run the program has gone to China!)

Brilliant or stupid are the two modes of operation for far too many classical systems; one is the most easily demonstrated and the other is the most usually presented (you figure it out).

The 'always' answer might stem from the observation that the only significant AI systems to date are classical systems; the connectionists may lay claim to the possibility of non-brittle AI systems but they have yet to deliver a substantial existence proof. Nevertheless, the CP advocates argue, as point 4 in Waltz's list and Smolensky's "soft constraints", that connnectionist systems will degrade gracefully in sharp contrast to the graceless behavior typical of classical systems. And there is some hardish evidence to support the claim (e.g. [Smolensky, 1986]).

"But", it is argued by supporters of classical AI, "it may be true that AI models have tended to be brittle. Nevertheless, it is not a *necessary* consequence of the classical approach that they should be so, see, for example, [Quinlan, 1987] – an implementation of non-brittle decision trees. And see [Kearns et al., 1987], for a formal theory of approximate correctness. Both are classical."

[Fodor & Pylyshyn 1988], for example, hold the view that brittleness is no more a necessary characteristic of classical systems than it is of connectionistic ones. This standpoint is in fact the key to much of their attempted demolition of the CP. They claim that the two defects in the usual arguments for preferring connectionist systems are:

1. The objections depend on properties that are not intrinsic to classical systems. "There can be perfectly natural Classical models that don't exhibit the objectionable features."

2. "The objections are true of Classical architectures insofar as they are implemented on current computers, but need not be true of such architectures when differently (e.g. neurally) implemented. They are, in other words, directed at the implementation level rather than the cognitive level." Fodor & Pylyshyn spend some time on the question of what it is to be an explanation *at the cognitive level.* The cognitive level is specified by those states of the system that encode properties of the world. The elements of most connectionistic explanations are below this level. They emphasize the way that the cognitive level states can be implemented. This is a summary, hopefully not a travesty, of their point about levels of explanation. Needless to say, the connectionists do not agree.

Clearly not all the news is good news, as we've just seen. It's time to explore the dark side of the paradigm.

## 4.7 Subsymbolic Connectionism: the Real News

In conventional software engineering the comprehensibility of programs is of the utmost importance to support both tracking down known errors, and alteration of the system in response to changing requirements. In fact, comprehensibility of representations is important throughout the design and implementation stages as well. In AI the necessity for readily comprehensible systems takes a huge leap in importance (a claim that I have underpinned in the earlier sections): most AI system development is, of necessity, exploratory, and the maintenance of dynamic (given the presence of machine learning) AI systems puts even stronger demands on understandability than does the analogous task associated with static, conventional software systems.

## 4.8 Reasoning with Amorphous Complexity

The less optimistic view of the potential of connectionist architectures in AI hinges on the observation that they have a disconcerting similarity to brain structure: both subsymbolic networks and brains are conceptually opaque. What the observer's eye tells the observer's brain about neuroanatomy is that it is very complicated and embodies no underlying plan, apparently (at least no plan relating to the cognitive behavior that it exhibits). The apparent chaos makes random connectivity look somewhat ordered by comparison. Subsymbolic connectionist networks of any reasonable size will present exactly the same unhelpful appearance of white noise to the curious programmer.

The very features that make subsymbolic networks attractive – distributed representation and the simple homogeneity of parallel activity transfer – also make them singularly incomprehensible representations to attempt to reason about. Why a certain model is doing what it is observed to be doing, and what would be needed to make it do something differently, are both extremely difficult questions to answer. They are also vitally important for the support of computational-model building as scientific research rather than as more-or-less random hacking.

[Dreyfus & Dreyfus 1990] add a further twist to the knife in this understandability problem. As they point out: not only do we face the difficulties of dispersal of our symbolic concepts within connectionistic networks, but in addition, a self-adaptive network may learn to exhibit intelligent behavior in some domain in terms of features that are not a combination of features that people would ordinarily use. They make the point that intuitive expertise may be reliant upon distributed representations of features that are totally alien to our rational analysis of the phenomenon under study.

Despite the interest and euphoria that connectionistic research is currently generating, in reality the demonstrations of implemented networks have been limited to very small and specific problems: analysis of a specific, very simple, electronic circuit; parsing a couple of specific sentences, etc. Now this small-scale beginning would not be so damning if it were not for the fact that these demonstration models are carefully hand-crafted for the specific demonstration in mind, and fail utterly on any other task. [Pollack 1986] raises this problem of static specificity in connectionist models and he proposes a scheme for dynamic reconfiguration of these networks. Certainly something along these lines is sorely needed. Pollack's proposal of "cascaded networks" in which one network (the "context network") sets the weights in another network (the "function network") remains to be comprehensively stated.

One possible rejoinder from the CP camp might be that sophisticated learning mechanisms will obviate (or at least minimize) the need for human intervention in the development of a CP-based system. Just set up the initial network, feed it with instructive examples, and machine learning will automatically deliver the final system. This possibility is taken up in a later section.

## 4.9 The Myth of Empirical Guidance

Another unhelpful visitation from reality comes in the form of the empirical constraints that flow from research in neuroanatomy. Figure 1 sketches the "sunny" view of modelling within the CP: information about brain structure and function provides constraints on the building blocks and the framework for interconnection, whilst observations of human behavior are used to reject or confirm aspects of overall system function.

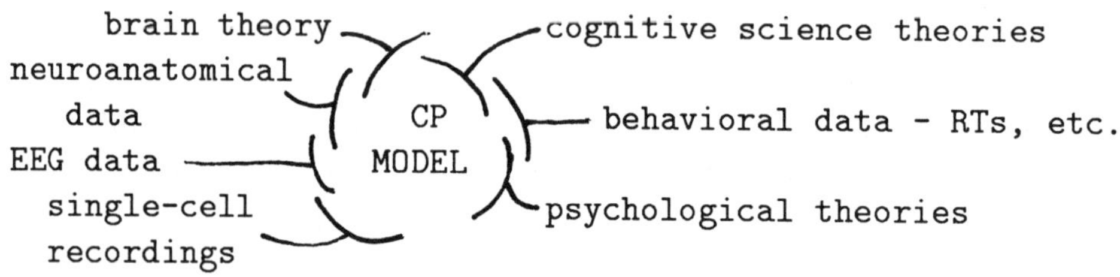


**Figure 1:** An optimistic view of CP-based modelling

In truth there is much data on the behavior of single neurons (single cell recordings) and also on the behavior of large collections of neurons *en masse* (as measured by, for example, EEGs), but there is very little beyond pure speculation that relates these two classes of data either to each other or to high-level cognitive behavior (vision is something of an exception).

Inferring the behavior of a brain, at the level of intelligent decision making or problem solving, from the evidence of single cell recordings (i.e. sticking a microprobe into individual neurons to record the rates at which they bleep) is like trying to shed light on the workings of a large power station by carefully recording, say, the fluctuations of a single valve whose interaction with the millions of other components is largely unknown; it's just like that, only much worse. Similarly, EEG data and the like provide some sort of 'averaged' measure of the brain's electrical activity. Drawing inferences from this class of measurement is like trying to infer the internal workings of an elephant from measurements of the surface temperature at certain selected locations on its skin. It is true that the temperature measurements vary in a predictable manner and that correlations between temperatures at specific places on the body and the organism's activities can be quite good. But how can we hope to move from this type of data to predictions or even explanations of what's going on inside the beast other than in terms of heat generating capabilities?

Consequently, there is the same paucity of firm inferences from empirical data pertaining to the behavior of moderately-sized neuron networks. In addition, in most neural models the experimenters are loath to equate their network nodes with neurons ([Feldman & Ballard, 1982], for example, suggest that their symbolic nodes represent "say 10" neurons); they are usually viewed as something more like Hebb's [Hebb 1949] cell assemblies – objects whose exact relationship to any observable anatomical features of the brain is never quite specified. So the reality of empirical constraints is somewhat thin for a few different reasons; the lack of empirical constraints for moderately-sized networks of neurons; inherent fundamental uncertainties about the significance and the role of neurons in our brains; and a failure to commit to any firm structural correspondences between connectionistic networks and brain anatomy. The mutually beneficial exchange of data is more of a promise of future riches when this, perhaps not awfully big, gap is bridged. It is a promise that not even the most wildly optimistic symbolic semantic network modeller is likely to see in his models, but it is only a promise for the future of CP-based work, nonetheless. Figure 2 is a realistic (or is it just pessimistic?) view of the context of CP-based modelling. The associations with brain theory are more a promise for the future than a reality, and the small-scale of current CP-based models does not typically generate behavior that can be compared with human behavior (although there are a few exceptions).

| | | |
|---|---|---|
| brain theory | | cognitive science theories |
| neuroanatomical data | CP | behavioral data, RTs, etc. |
| EEG data | MODEL | psychological models |
| single-cell recordings | | |

**Figure 2:** The realistic view of CP-based modelling

In, for example, the [Johnston et al. 1983] connectionist model of basic learning behaviors, one significant option in the mechanism for distributing activity out from a node concerned the relationship between the node's activity level and the number of outgoing links. In particular, should the amount of outgoing activity be divided between the outgoing links or duplicated for each link? Put another way: do the recipient nodes receive an amount of activity that is dependent only upon the source node's activity level and the weight of the link, or should it also depend upon the number of recipient nodes? The empirical data contained nothing to guide this decision. And even if some relevant information on brain function were known: how does knowledge of a multi-meganode network translate into constraints for a network model that is, of necessity, many orders of magnitude smaller? The strategy adopted, by default, was to explore the alternatives empirically – i.e., we used the suck-it-and-see procedure. We programmed up both alternatives, explored their consequences empirically, and adopted the one that worked best. The finding was that the behavioral data was modelled better sometimes by one function and sometimes by the other.

## 4.10 What's the Stopping Rule?

A further problematic aspect of the CP concerns questions of timing as it so often does when parallelism is countenanced. How do we know when a pattern of activity has emerged? Or how do we move the system on from a stable state? These are more or less opposite sides of the same problem. If the network is truly dynamic, fairly stable patterns of activation will periodically emerge and then break up again to contribute their activity to subsequent transient activation patterns. When do we attempt to "read off" some problem solution or decision or whatever?

Many current connectionist systems are one-shot constructions that are designed to head for a stable pattern of activation and stay there. Thus it is fairly clear when a problem solution or decision or whatever has been reached, but the problems then becomes how to move the system on to subsequent decisions etc.

Another view of this problem reveals it to be the problem of sequencing. In a parallel processing system, however massively parallel it is, there also has to be a sequencing aspect: at some level one phenomenon follows another. Nothing could be simpler in classical modelling. But the CP offers no obvious way to set up the sequencing of events.

[Waltz 1987] raises this general problem as well as several others including the problems of debugging and understanding connectionist systems. [Touretzky & Hinton 1986] added some "gates" to their CP-based model of a production system in order to achieve the necessary sequential ebb and flow of activity. [Sharkey 1989], for example, has explored several different means of producing appropriate sequences of events in his CP-based approach to natural language understanding.

Other attempts to introduce some serial ordering into CP-based models can be found in [Pollack 1987], [Elman 1988], [Gasser & Dyer 1988], and [Jordan 1986].

## 4.11 Single-Minded Models

The emphasis in the heading of this section is very much more on the "single" rather than the "minded" aspect of connectionistic models. They are painstakingly constructed to demonstrate one task, and one task only. This is not even one sort of task (e.g. sentence parsing rather than image processing). It is one very specific task (e.g. the network will parse one sentence, or perhaps a very small set of minor variants on the chosen exemplar).

[McCarthy 1988] provides us with just one example of this type of critique of connectionistic models. He calls it lack of *elaboration tolerance* which is a lack of "the ability of a representation to be elaborated to take additional phenomena into account." In response, Smolensky ([Smolensky 1988], the Author's Response section) admits that McCarthy's charge is right on target. He continues: "At this point, connectionist models tend to be developed with the barest minimum of representational power for the target task. If the task is beefed up even a little, the model has to be scrapped." He then adds that this weakness is not inherent in CP-based models. It is just a phenomenon that we are observing at the dawn of connectionism; given time, it will, presumably, disappear.

[Pollack 1989] addresses the issue of representational adequacy, and he also has something to say on the lack-of-elaboration problem. He claims that the [Rumelhart & McClelland 1986] model of past-tense formation in English (e.g. "kill" becomes "killed" but "go" becomes "went" except for young children who erroneously say "goed" until it's beaten out of them) cannot be switched from, say, past to present tense without another 200,000 weighted links. And the travelling salesman network of [Hopfield & Tank 1985] must be entirely reconfigured if we want to add one more city to the problem.

Like so many of the imperfections of this shiny, new paradigm (if that's what it is), it remains to be seen whether intolerance to elaboration is a superficial scratch that will be polished out, or whether it is a deep gouge that the modellers will have to live with as best they can. But it clearly compounds the incremental development problem as it is closely intertwined with the conceptual opacity feature.

That's quite enough on the pros and cons of connectionism. It might be argued that many of the deficiencies of the CP appear to be weaknesses in the paradigm only because we are viewing the CP from a classical perspective. Thus the fundamental problem of conceptualization that is necessary to support incremental development will be circumvented (rather than solved) by CP-based models. Connectionist models will *learn* to cxhibit the desired behavioural characteristics. The need to constantly delve into the innards of the networks will largely vanish.

## 4.12 The "Learning" Algorithms are only Training Regimes

The conceptual opacity of connectionist systems is typically not denied but it is also not thought to be a major problem because even at this early date, close to its inception, powerful and provably convergent learning algorithms are already apparent. So we don't have to comprehend all the nitty-gritty of our distributed subsymbolic networks in an effort to incrementally upgrade their performance in the time-honoured AI manner. We just expose them to a rich set of examples and they'll learn to do things right – they'll automatically upgrade themselves.

This is just another example of the fact that connectionism is indeed a new paradigm, and that the old AI tradition of hand-crafted, incremental system development is (like punched card input) a thing of the past. Or is it? How powerful are connectionistic learning algorithms –

back propagation, simulated annealing, etc.?

Well, they are powerful (even if they are monstrously time-consuming), but *only if we know the right answers* to the problems that they are going to learn about. Open-ended learning is no more a solved problem in connectionism than it is in classical AI. It is true that there are some examples of connectionist systems generalizing, i.e., having learned something beyond the information explicitly in the training set. Hinton's (1989) connectionist system that learns family relationships is trained on 100 of the possible 104 examples and successfully generalizes to the remaining four.

A related charge that is sometimes leveled at such self-adaptive connectionist systems is that they are just elaborate curve-fitting techniques. To call a central feature of connectionist models "just elaborate curve-fitting" doesn't sound too good, but it may not, in actuality, be too bad – if true. After all, Shakespeare's works are just words, and we are primarily water with a few other elements mixed in. To return to the original slur: it is conceivable that human intelligence is "just elaborate curve-fitting", I suppose. If the necessary degree of elaboration requires the use of heuristics and the consequent loss of certainty that simple curve-fitting algorithms (such as the least squares algorithm) carry with them, then I could believe that intelligence is just elaborate curve-fitting.

The lack of elaboration tolerance typical of CP-based systems further weakens the case for connectionist learning algorithms. The systems can be trained to exhibit only a very narrow functionality. To get them to do something a little bit different seems to require *starting from scratch.* Thus the NETtalk system for learning to pronounce English is quite impressive: as example text is run through the system and the link weights are adjusted in the manner dictated by the back propagation algorithm, the initial random babbling gradually changes to alien, but quite understandable, pronunciation, and then with further training it quite quickly becomes quite good pronunciation. But if we wish to develop a system for pronouncing English with, say, a Welsh accent then we must go through exactly the same long training process. There is no known way to identify, and hence modify or substitute, components of the network (i.e., subnets) that are responsible for any of the essential differences between the "standard" and the Welsh pronunciations of English.

The expected response from connectionists is that much of my criticism is valid but that it is early days yet, and better results will be forthcoming, given a little time. This riposte has some merit, perhaps, but such a belief can only be based on blind faith.

In conclusion, there are some general points about self-adaptivity in CP-based systems:

(a) Mechanisms for machine learning are one of the particularly bright spots in the CP, but

(b) Current mechanisms have limitations that fix them way below the necessary level of sophistication, and

(c) Such a "black box" methodology could conceivably be an acceptable style for engineering practical systems; but could it be part of a "science"?

Let me now expand our scope of interest just a little. It may be that the difficulties that arise in connectionist modelling are not just a parochial concern to aspiring AI scientists, these problems may also illuminate some general issues of the impact of computers on the nature of classical science.

## 4.13 Complexity and the Failure of Classical Science

The triumph of physics is that it discovered the relatively simple laws of nature, and this is the challenge for the science of AI with respect to intelligence. Or is it the luck of physics that nature can be approximated quite well by relatively simple laws, and the curse on AI that intelligence is not a similar sort of phenomenon?

Classical science (as epitomised by physics) thrives in a milieu of low complexity: there are few significant variables and their interrelationships are very limited (at least, models based on these sorts of assumptions are quite good approximations to reality). Thus the behaviour of gases, for example, can be modelled quite accurately by consideration of only pressure, volume, temperature, and a simple relationship between these three quantities. This is a domain of low complexity.

In the mathematical sciences good practice is exemplified by the construction and application of formal proofs. It is widely believed that the science of programming should be founded on similar notions of proof and verification. So scientific AI with its goal of furthering our understanding of the nature of intelligence, is necessarily involved with empirical procedures (because intelligence is not an abstract notion but a real-world phenomenon) like the classical sciences, and is also similarly allied to the mathematical sciences in that its prime mode of exploration and enquiry is the formal activity, programming. We might then hope that a science of AI could harness the best that each of these two forms of scientific endeavour has to offer.

Let's now look at the mathematical sciences. A good proof in the more mathematical sciences is simple and elegant. It is readily comprehensible (to the mathematically minded) both at the step-by-step level of formal details, and, more importantly, at a general level of the direction and style of the proof. To communicate a new proof effectively we have to be able to capture its "essence" and innovative nature in terms of accepted generalities such as by analogy with older and widely-accepted proofs. Thus the innovative computer proof of the four-colour conjecture is a scheme whose status as a proof (or a good proof) is a hotly debated issue. And the root cause of the general dissatisfaction with this proof was its complexity. It was too complicated; it lacked conceptual transparency. A typical comment from its detractors is that study of this proof does not lead to significant theoretical insights into the original problem. It is a proof comprised of many individual special cases rather than a few abstract generalizations. It is thus a complex proof but with a complexity founded upon a diversity of special cases (and not, for example, upon a long chain of reasoning based upon just a few abstract generalizations). Finally, it is a proof that was only possible with the aid of a computer.

The computer lures us into domains of unmanageable complexity. It tempts us to explore worlds wherein the techniques of classical science are quite powerless. It seduces us into the construction of theories of totally incomprehensible complexity.

Large and complex programs are the accepted currency of AI (despite much complaint about the crucial role that these constructions play). Supposed theories (of say natural language processing, or even scientific discovery) are communicated in a concise and comprehensible manner, either in prose alone, or in prose punctuated with formal symbolic expressions. But can these "theories" be tested? Certainly not with anything like the precision with which theories in the hard sciences can be tested. Why not? The reason is that the comprehensible theoretical statements are such gross simplifications of the original programs within which they were developed that a myriad of crucial theoretical detail is missing when we come to construct a new program from the "theory" (i.e., run experiments to corroborate or refute it).

We just don't know how to deal "scientifically" with theories of the complexity that we can develop using a computer. What's the purpose of this digression? Well, the computer, which at first sight has just provided the scientist with a fast and accurate substitute for pencil and paper, has in fact opened up a whole new world for the inquisitive researcher to explore – the quantitative increase in speed and accuracy has led to a qualitative change in complexity. Thus in the example of the "proof" of the four-colour conjecture, its status as a good proof is undermined by the sheer number of special cases that the computer exhaustively considered. The notion of special cases (i.e. dividing a domain into a number of independent possibilities and then providing independent proofs for each case) is not itself contrary to the idea of a good proof. It was the quantitative leap in the number of special cases that comprised the four-colour proof that blackballed this "proof". And, to finally get to the point, certain ways of using computers prolong the effective use of tools from classical science in domains of increasing theoretical complexity, whereas others, such as connectionism, do not. So, to my mind, connectionism aggravates the already difficult problem of complexity in AI, and thus threatens to close the door on theoretical insights into our problems earlier than more traditional approaches to the construction of programmed models – unless we can "beef up" our notions of, for example, the experimental testing of theories.

In classical science, we can generally get away with sloppy notions of repeatability, for example (although the recent cold-fusion saga has highlighted the ill-defined nature of repeatability even in the hardest of classical sciences). The important variables are few, and the rest of the "variables" in the world can be neglected without noticeable penalty for most of the time. In the mathematical sciences, it's the prose in the proof (and not the symbolic expressions) that carry the weight of the argument. It is, for example, often the word "similarly" wherein much of the interesting information is packed in a proof. The whole formal-proof notion is riddled with vagaries, but we can largely ignore them and still make significant progress piling successive proofs one on top of another. But in AI, which has to deal in highly-detailed, complex theories explored and developed with the uncompromising logic of computers, the relatively minor perturbations in the performance of the classical sciences (both empirical and mathematical) become combined and enlarged to the extent that, far from being ignorable to a first approximation, they tend to dominate the activities, and undermine the whole endeavour.

But if it's not theoretical insights we're after, then the story may be different, especially if powerful learning algorithms are available. We set up a basic network, teach and train it appropriately, and end up with a system that performs just as we had hoped. Its workings are a total mystery, but it performs a useful task with high reliability, and so it is a well-engineered product. And thus connectionism will be judged to be an effective new paradigm for engineering AI systems. [Sejnowski and Rosenberg's 1986] NETtalk (briefly described earlier) is a good example here.

It is easy enough to be negative, so I'll finish off with a few suggestions of a positive nature.

## 4.14 Possible Routes to a Science of AI

Connectionist networks have many attractive qualities; however, they are also rife with basic problems. The perceptual opacity of these networks does not bode well for the prospect of solving the problems by inspection and intuitive reasoning. Three approaches that may eventually contribute to the understandability of neural architectures are outlined below.

## 4.15 Formal Analysis

Formal analysis seems to be one promising route to the ready conceptualization of subsymbolic networks. One popular approach to formal analysis is to apply the tools of thermodynamics (the nodes are perhaps analogous to molecules and the arc weights to the intermolecular energy). Thus we have the Boltzmann machine approach in which noise is injected into the network to overcome local minima followed by simulated annealing to discover minimum energy states of a network (see [Ackley et al., 1985]). A closely related style of CP-based system is the "Harmony theory" models of [Smolensky 1986], the formal treatment of which is based upon work in statistical mechanics.

Both of these formal schemes are more than a method of analysis. They are also a methodology for problem solving with these networks. It is a general methodology that works quite nicely on a number of examples, but its advocates have yet to integrate these one-time-stabilisation networks into a continuous sequence of events. Currently, we have a number of one-shot, hand-crafted, special-purpose problem solving networks. Use of these networks to reproduce extended intelligent behavior, which involves a sequence of stabilisation and destabilisations to solve a connected string of subproblems, has yet to be realized.

Another, as yet largely untried, route to gaining insights through formal analysis might be to apply the power of graph theory. [Hage & Harary 1983], for example, have demonstrated how the concepts, theorems, and techniques of graph theory can be used to systematize and explain ordinary ethnographic data. It is thus reasonable to expect knowledge networks with their fundamentally structural representations to be even more amenable to graph theoretic analysis. One application of graph theory in this general area is provided by [Nicole, Lloyd & Ward 1988].

Graph-theoretical analyses of knowledge representation schemes appear to be scarce, if not non-existent. Classical semantic networks, with their perceptual clarity and heterogeneous structure, would not be expected to encourage such analyses: there is both a lack of need, and an apparent difficulty associated with anything more than a superficial analysis. But connectionist networks on the other hand appear to offer the reverse characteristics: a pressing need for analytical insights, and a representation that invites formal analysis.

It appears that there may be two somewhat different types of graph-theoretic analyses that are suggested by current problems with connectionist networks: analyses of the scope and limitations of proposed functions (arc weight adjustment as a result of behavior; accumulation of activity in nodes; activity transfer; etc.) *vis-á-vis* different classes of networks (e.g., bipartite graphs or multilayer networks); and secondly, as a tool to promote understanding of the behavior of given networks and associated functions (i.e., provide insight into why a network is doing what it is observed to be doing).

Advocates of CP-based systems make much of the "dynamical" nature of this paradigm. Graph theory appears to be a tool for analyzing static structures, but it is not obvious to me that this apparent mismatch is a real barrier to graph-theoretic analyses. [Pollack 1989], incidently, opts for "connectionist fractal semantics" as a future boom area based on the fractal geometry of [Mandelbrot 1982].

## 4.16 Software Support Systems

A rather different route to increasing the understandability of subsymbolic networks is to build and apply software support systems: programs whose function is to interpret some aspect of the network system and present it in a more understandable form to the human user. The P3 system of [Zipser & Rabin 1986] is one such system that is designed explicitly for connectionistic research.

A second possibility in this class of support systems is the "software oscilloscope" of [Nieper & Boecker 1985]. They describe a software equivalent of the electrical engineer's oscilloscope: a program that is 'connected to' another program, whose behaviour we wish to explore, in order to observe certain functional characteristics of this latter program, characteristics that would otherwise be invisible to the interested observer. This idea was not developed, and has not been applied, to subsymbolic networks, but I see no reason why it should not yield useful results in this new domain.

The software oscilloscope is just one example from the newish Computer Science subarea called Program Visualization which is the use of graphics to illustrate some aspect of the program or its run-time execution (see [Myers, 1988], for an introduction to this realm).

## 4.17 Approximate Translation – the Truth About Mendacity

This last offering is not so much another, and different, route to enhancing the understandability of network models (especially subsymbolic ones), it is more a suggestion as to how we might most usefully apply both formal analytical techniques and software support systems.

One style of interpreting a mass of fine detail so that it becomes more understandable is to generate high-level approximations – i.e., to tell less than the whole truth (just like I've been doing throughout this chapter). There's lies, there's damn lies, and then there's approximations to the truth; it's this latter category that we want to focus on. Thus a pattern of activation in the subsymbolic network may be approximated by a single symbolic node in a much more comprehensible, but not totally accurate, network of symbolic nodes. [Smolensky 1987] makes the point that "symbol-level structures provide only approximate accounts of cognition, useful for description but not necessarily for constructing detailed formal models." And, I would add, perhaps necessary for supporting attempts to comprehend these detailed formal models in order that they may be usable.

A vague illustration of this possibility is provided by Figure 3. This general idea is allied to the Fodor and Pylyshyn criticism that connectionist systems lack compositional semantics. My point is that, with a little fibbing, such systems may be effectively visualizable *as if* they had such compositional structure at a useful conceptual level.

[Michie & Johnston 1984] also suggest the need for such interpretations, but in the context of making expert systems more understandable.

The truth about the need for mendacity is that it is an essential one when faced with the complexity barriers of AI: design, development, communication of both problems and solutions, and comprehension in general are all issues that will benefit from the judicious use of approximations to the truth. This is not, of course, to be taken as *carte blanche* for flagrant lying about achievements on all fronts. Something much less pernicious than this does much damage in presentations and publications of AI already. Nevertheless, there is, I am sure, a need for accepted, well-defined schemes of approximate representation of complex software systems; one such scheme, which I called stepwise abstraction, is detailed elsewhere (see [Partridge, 1986], for the exhaustive treatment with respect to AI software, and [Partridge, 1990], for a terse example).

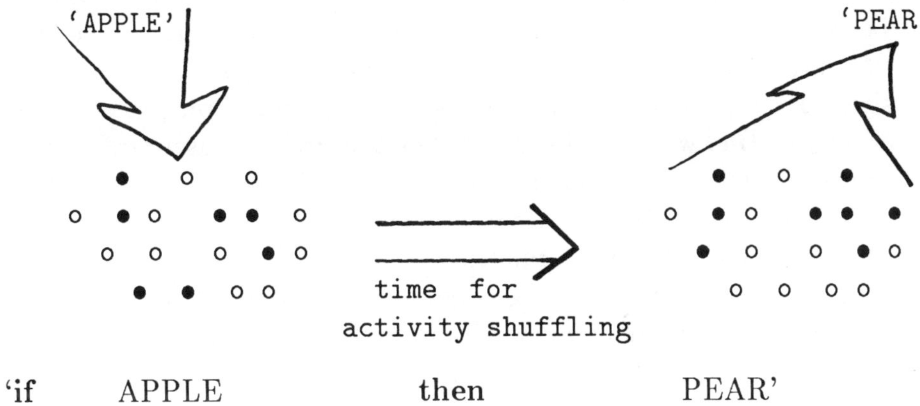


**Figure 3:** An approximate symbolic macroinference from a collection of microinferences

In sum, connectionist architectures, of all varieties, hold much promise for the realization of AI. They also present a number of difficult, unsolved problems, but above all they threaten to be useless unless we can find ways to render them comprehensible to humans.

## 4.18 In Conclusion

I have argued that connectionism looks to be bad paradigm for the furtherance of a science of AI, and that it is primarily an understandability problem that arises from one of excessive complexity. But it is not just a problem of complexity, but complexity intertwined with the problem of separating theory-irrelevant details from theoretically significant ones in our experimental programs. Thus in a 4,000 line LISP program we might strip out 1,000 lines of theoretically non-significant, but operationally crucial code. We are then still left with a 3,000 line function that is our theory. This is the excessive complexity point.

The compounding problem is how do we perform the separation in the first place? What is the decision mechanism? What justification can we provide for the details that we abandon as not relevant, and for the structure that we present as the theory? I don't think that we have, as yet, any good answers to these questions. But if the answer requires some degree of isomorphism between program and theory, then CP-based models are less promising than classical ones. It is true that this demarcation problem applies to the pursuance of a science of AI, classical or connectionistic, if the science is founded on programs as experiments. But with CP-based models the problem assumes a more extreme form, for the basic structure of CP-based models is even further removed from the structure of comprehensible theories than the basic structures typical of classical models.

[Chandrasekaran's 1990] argument for an information-processing (IP) analysis of phenomena of interest as a way of separating out the essential elements of the theory from implementation-level commitments, provides a much-needed concretization of my arguments. According to Chandrasekaran, many connectionistic models incorporate their own versions of modularity and composition. And the resultant components of the models are, in effect, implementations of IP abstractions. From our current perspective, we can say that a need to comprehend connectionistic models (which includes their initial construction) has resulted in many of them being, not manifestations of a totally new paradigm, but novel implementations of the components of a thoroughly conventional (and thus relatively comprehensible) theory – for Chandrasekaran this is an IP theory.

Just suppose that, contrary to most current indications, we discover that many aspects of human intelligence can be accounted for by a few simple laws of activity transfer, of link-weight adaptation, and of activity accumulation in recipient nodes. If this were the case, we could, presumably, build vast networks (I assume that the networks would *have* to be very large) within which each node and link obeyed the known laws, and these models would behave intelligently. But would we be very much the wiser about the nature of human intelligence? I think not.

## References


Ackley, D.H., Hinton, G.E., & Sejnowski, T.J., [1985]: "A learning algorithm for Boltzmann machines", *Cognitive Science*, vol. 9, no. 1, 147-169.

Chandrasekaran, B., [1990]: "What kind of information processing is intelligence?" in Partridge, D., & Wilks, Y., (eds.), *The Foundations of AI: a sourcebook*, 14-46, Cambridge Univ. Press. UK.

Dreyfus, S.E., & Dreyfus, H.L., [1990]: "Towards a Reconciliation of Phenomenology and AI", in Partridge, D., & Wilks, Y., (eds.), *The Foundations of AI: a sourcebook*, 369-410, Cambridge Univ. Press. UK.

Elman, J.L., [1988]: "Finding structure in time", Rep. 8801, Center for Research in Language, UCSD, San Diego, CA.

Feldman, J.A., & Ballard, D.H., [1982]: "Connectionist models and their properties", *Cognitive Science*, vol. 6, 205-254.

Fodor, J.A., & Pylyshyn, Z.W., [1988]: "Connectionism and cognitive architecture: A critical analysis", *Cognition vol. 28*, 3-71.

Gasser, M., & Dyer, M.G., [1988]: "Sequencing in a connectionist model of language processing", *Procs. 12th Internat. Conf. on Computational Linguistics*, Budapest.

Hage, P., & Harary, F.N.,[1983]: *Structural models in anthropology*, Cambridge Univ. Press: Cambridge, UK.

Hebb, D.O., [1949]: *Organization of Behavior*, Wiley: NY.

Hinton, G.E., [1990]: "Connectionist learning procedures", *Artificial Intelligence*, vol. 40, nos. 1-3, 185-234.

Hinton, G.E., [1986]: "Learning in massively parallel nets" (extended abstract only), *Procs. AAAI-86*, Philadelphia, 1149.

Hopfield, J.J., & Tank, D.W., [1985] " "Neural" Computation of Decisions in Optimization Problems", *Biological Cybernetics 52*, 141-152.

Johnston, V.S., Partridge, D., & Lopez, P.D., [1983]: "A Neural Theory of Cognitive Development", *J. Theor. Biol.*, 100, 485-509.

Jordan, M.I., [1986]: *Serial order: a parallel distributed processing approach*, ICS Rep. 8608, Institute for Cognitive Science, UCSD, La Jolla, CA.

Kearns, M., Li, M., Pitt, L., & Valiant, L.G., [1987]: "Recent results on Boolean concept learning", *Procs. 4th Internat. Workshop on Machine Learning*, 337-352, Irvine, CA.

Mandelbrot, B.B., [1982]: *The Fractal Geometry of Nature*, Freeman. San Francisco.

McCarthy, J., [1988]: "Epistemological challenges for connectionism", *Behavioral and Brain Sciences*, vol. 11, no. 1, p. 44.

Michie, D., & Johnston, R., [1984]: *The Creative Computer*, Penguin. UK.

Myers, B.A., [1988]: "The state of the art in visual programming and program visualization", in Kilgour, A.C., & Earnshaw, R.A., (eds.), *Graphics Tools for Software Engineering*, The British Computer Soc. Internat. State of the Art Seminar, March 16th, 24 pages.

Nicole, D.A., Lloyd, E.K., & Ward, J.S., [1988]: "Switching Networks for Transputer Links", *Procs. 8th Tech. Meeting Occam User Group*, Sheffield, UK, March, 147-165.

Nieper, H., & Boecker, H.D., [1985]: *Making the Invisible Visible*, Dept. CS, Univ. Colorado, Boulder.

Partridge, D., [1986]: *AI: applications in the future of software engineering*, Ellis Horwood/Wiley: Chichester, UK.

Partridge, D., [1990]: "What's in an AI program?", in Partridge, D., & Wilks, Y., (eds.) *The Foundations of AI: A sourcebook*, 112-118, Cambridge Univ. Press: Cambridge, UK,

Partridge, D., & Wilks, Y., [1987]: "Does AI have a methodology which is different from software engineering?", *AI Review, 1, 2*, 111-120.

Pollack, J.B., [1986]: "Universal Neural Networks", *Proc. New Mexico CS Conf.*, Las Cruces, NM.

Pollack, J.B., [1987]: *On Connectionist Models of Natural Language Processing*, PhD Thesis, Dept. C.S., Univ. Illinois at Urbana- Champaign.

Pollack, J.B., [1987a]: "Cascaded back propagation on dynamic connectionist networks", *Procs. 9th Conf. of the Cognitive Science Society*, Seattle, WA, pp. 391-404.

Pollack, J.B., [1989]: Connectionism: past, present, and future, AI Review, vol. 3, no. 1, pp. 3-22.

Quinlan, J.R., [1987]: "Decision Trees as Probabilistic Classifiers", *Procs. 4th Internat. Workshop on Machine Learning*, Irvine, CA, 31-37.

Rumelhart, D.E., & McClelland, J.E., (eds.) [1986]: *Parallel Distributed Processing: Explorations in the Microstructure of Cognition*, vols. I & II, MIT Press: Cambridge, Mass.

Sejnowski, T.J., & Rosenberg, C.R., [1986]: *NETtalk: a parallel network that learns to read aloud*, Tech. Rep. JHU/EECS-86/01, Dept. E. E. & C.S., Johns Hopkins University, MD.

Sharkey, N.E., [1989] "A PDP learning approach to natural language understanding", in Aleksander, I., (ed.), *Neural computing architectures*, 92-116, Kogan Page. London.

Smolensky, P., [1986]: "Formal modeling of subsymbolic processes: an introduction to harmony theory", in Sharkey, N.E., (ed.), *Advances in Cognitive Science 1*, 204-235, Ellis Horwood. Chichester, UK.

Smolensky, P., [1987]: "Connectionist AI, Symbolic AI, and the Brain", *AI Review 1, 2*, 95-109.

Smolensky, P., [1988]: "On the Proper Treatment of Connectionism", *Behavioral and Brain Sciences, vol. 11, no. 1*, 1-23, 59-74.

Touretzky, D.S., & Hinton, G.E., [1986]: *A Distributed Connectionist Production System, CMU-CS-86-172*, C.S. Dept., Carnegie Mellon University, Schenley Park, Pittsburgh.

Waltz, D.L., [1987]: "Connectionist models: Not just a notational variant, not a panacea", *Procs. TINLAP 3*, Computing Research Lab., New Mexico State Univ.

Zipser, D., & Rabin, D., [1986]: "P3: A Parallel Network Simulating System", in Rumelhart, D.E., & McClelland, J.L., (eds.) Parallel Distributed Processing, 488-506, MIT Press. Cambridge, Ma.

# Chapter 5

# Rules and Competence in Connectionist Networks

Martin Davies
*Philosophy Department, Birkbeck College, Malet Street, London WC1E 7HX*

It is often said that what is distinctive of connectionist networks is that they perform cognitive tasks without having knowledge of rules. Thus Rumelhart and McClelland ([1986], p.218):

> We would . . . suggest that parallel distributed processing models may provide a mechanism sufficient to capture lawful behaviour, without requiring the postulation of explicit but inaccessible rules.

And Norman ([1986], p.536):

> [A]lthough the system develops neither rules of classification nor generalizations, it acts as if it had these rules. . . . It is a system that exhibits intelligence and logic, yet that nowhere has explicit rules of intelligence or logic.

Since competence is usually identified with knowledge of rules (Chomsky [1965], p.8), we might expect it to be widely agreed that there is no such thing as connectionist competence. Yet Smolensky says ([1988], p.3):

> It is likely that the view of the competence/performance distinction that arises from the connectionist approach will successfully heal a deep and ancient rift in the science and philosophy of mind.

So what is this notion of competence that is applicable to connectionist networks? And how is it related to knowledge of rules?

A broad strategy for approaching these questions is fairly clear. Rumelhart and McClelland [1986] are operating with a particular conception of knowledge of rules: 'the *explicit inaccessible rule* view' (p.217). If there is to be such a thing as connectionist competence, then some other conception of knowledge of rules must also be in play.

Our search for connectionist competence will take us through several different notions of knowledge of rules. But let us start with the way that rules themselves figure in the analysis of cognitive tasks.

## 5.1 Cognitive Tasks and Their Analysis

We can distinguish, very roughly, two kinds of cognitive task. There are those that involve a transition from information to information, and those that involve moving from information to action. Let us focus upon tasks of just the first kind.

The analysis of a cognitive task may reveal that there are certain rules or principles which would be adequate to mediate between the information that is available as the input to the task and the information that is delivered as the output of the task. This is to say that the information delivered as output can be inferred from the information available as input, given the information that is embodied in the rules or principles.

In the simplest case, in which the inference in question is deduction, a task analysis can then be cast in the form of a deductive theory. The discovered rules or principles of the task domain are cast as non-logical (proper) axioms of the theory, in such a way that from premises expressing input information conclusions expressing the corresponding output information can be derived (using some background logic).

## *An example*

Consider the cognitive task of understanding visually presented sentences. This task involves a transition from information about what sentence has been presented, to information about what message has been conveyed. In order to keep the example simple, we shall ignore all the complexities of natural language syntax and of the context dependence of interpretation.

The sentences to be considered are just the hundred sentences that can be composed from ten one-place predicates, 'F', 'G', ... , 'O', and ten names 'a', 'b', ... , 'j'. These sentences have very simple meanings. For example, the sentence 'Fa' means that John is bald, while the sentence 'Fb' means that Harry is bald. The cognitive task of understanding these sentences involves, for example, moving from the information that what was presented was 'Fa' to the information that what was conveyed was that John is bald. (For this example, see Evans [1981].)

The formal semantic business of constructing a systematic theory of meaning for these sentences amounts to a task analysis for this imaginary cognitive task.

A theory of meaning is quite literally a deductive theory, and the proper axioms of such a theory specify the contributions made to sentence meanings by the ten predicates and the ten names. Thus, for example, one of the axioms of a theory of meaning may tell us, in effect, that if what is presented is any sentence containing the predicate 'F' (whether it be 'Fa', or 'Fb', or ... ), then what is conveyed concerns some particular person (whether it be John, or Harry, or ... ) to the effect that the person is bald. Likewise, another axiom may tell us that if what is presented is any sentence containing the name 'a' (whether it be 'Fa', or 'Ga', or ... ), then what is conveyed concerns John to the effect that he has some particular property (whether it be that he is bald, or that he is happy, or ... ).

As this example illustrates, a theory of a task domain does very much more than specify what the cognitive task is. Certainly, it does more than merely specify an input-output function in extension. On the other hand, such a theory is not, by itself, an account of the operation of any particular cognitive system. If the theory is to figure in the explanation of a system's behaviour, then there must be some relationship between the system and the rules articulated by the theory. The term "knowledge of rules" serves to label such a relationship.

Given this relationship between system and rules, the theory that articulates the rules serves to characterise the competence of the system: a theory of the task domain becomes a theory of competence. But just what is the explanatory role of a competence theory? This depends upon the nature of the relationship between the system and the rules. There are at least as many conceptions of competence - and of the competence/performance distinction - as there are notions of knowledge of rules.

One possible kind of relationship between a physical system and the rules of a task analysis is captured by the explicit inaccessible rule view. It can be illustrated as follows.

## 5.2 Explicit Rules

The physical system has standing states which constitute linguistic encodings of the non-logical axioms of the deductive theory of the task domain; and it has transient input states which encode, in the same linguistic format, the available input information. In addition, there are processes which operate in the system to produce output states from input states and standing states in a way that mirrors the derivation of conclusions from premises and axioms in the deductive theory. In effect, these processes are the system's realisation of the background logic of the deductive theory; we could imagine them as the operations of a little man performing logical derivations on a blackboard.

There is a clear sense in which this physical system has knowledge of the rules of the task analysis; for all the rules that concern the task domain are explicitly represented in the machine.

A significant feature of this maximally explicit machine is that the states of knowledge of the proprietary rules of the task domain fall very far short of providing an explanation of the system's outputs given particular inputs. Those states would be impotent but for the logical processes, just as in the deductive theory the derivation of conclusions from premises requires the background logical rules in addition to the proper axioms.

As well as logical processes, the maximally explicit machine also requires processes to locate the pieces of task domain knowledge that are needed, given any particular input. The involvement of these search processes plays a crucial role in explaining, for example, reaction time differences in the system's operation upon different inputs, even in cases where the deductive complexity of the derivations of the respective outputs is the same. For it may be that the axioms needed are represented earlier or later in the sequence of knowledge states through which the search proceeds.

This explanatory gap between having knowledge of rules and using that knowledge provides the space that is needed for a competence/performance distinction. There are states of the system that embody knowledge of rules; and those rules are logically adequate to mediate between input and output information. But that logical fact will not make the machine work; and the states of knowledge of the rules are, by themselves, causally impotent. The operation of the machine depends upon processes of search and access, and processes of inferential transition. And those processes may be restricted in their application by limitations of time and of working memory, for example. Thus the performance of the system may fall short of full conformity to the rules, knowledge of which constitutes the system's competence.

It might be that, under ideal conditions for the search and transition processes, the system's performance would live up to its competence. But it does not follow that competence is simply idealised performance. On the contrary, on the explicit representation view, competence figures in the causal explanation of performance. The states of knowledge of the rules are implicated in the production of output states from given input states.

The wide explanatory gap between having knowledge and using it is not a feature of typical connectionist networks. In connectionist systems (McClelland, Rumelhart and Hinton [1986], p. 32):

> The representation of the knowledge is set up in such a way that the knowledge necessarily influences the course of processing. Using knowledge in processing is no longer a

> matter of finding the relevant information in memory and bringing it to bear; it is part and parcel of the processing itself.

This vivid difference is a symptom of the fact that connectionist networks perform cognitive tasks without having knowledge of rules in the sense favoured by the explicit inaccessible rule view.

We have here a major bifurcation. Explicit rules provide one account of competence and of the competence/performance distinction. Other accounts locate competence as "part and parcel of the processing". Searching for connectionist competence we have to follow the latter path rather than the former.

However, it would be wrong to represent connectionism's divergence from the classical symbol manipulation paradigm as turning solely upon the absence of explicit rules. For, as Fodor has stressed many times, symbol manipulation does not require explicit representation of the rules in accordance with which the symbols are manipulated. (And this cannot, in general, be required: the Lewis Carroll point.) What symbol manipulation does require is syntactic articulation in the transient input states. (See Fodor [1985], p.95, [1987], p.25; Fodor and Pylyshyn [1988], pp.59-61.)

What are implicit rules; and how does a system with implicit rules differ from a system that merely conforms to rules?

## 5.3 Implicit Rules

Fodor and Pylyshyn [1988] say (p.60): "Classical machines can be *rule implicit* . . . ". This suggests the need for a notion of knowledge of rules that does not require explicit representation. On the other hand, if the correlative notion of competence is to figure in causal explanation - and not play a merely descriptive role - then knowledge of rules must require more than just achieving the right input-output function. We can develop an intermediate notion using the idea of causal common factors.

In the maximally explicit machine, the states encoding the axioms play a causal role in the production of output states from input states;

that role mirrors the role played by the axioms themselves in the derivation of conclusions from premises. An axiom may be a derivational common factor in the deductions from two different premises of the respective conclusions about the task domain. In our example, the axiom that tells us the semantic contribution made to sentences by the predicate 'F' figures *inter alia* as a common factor in the derivations of interpretations for 'Fa' and for 'Fb'.

So, in the machine, the state encoding the axiom will be a causal common factor in the two processes that lead from input states corresponding to the two premises to the respective output states.

Suppose the machine is performing the sentence interpretation task. Then the encoding of the axiom about the predicate 'F' will figure as a causal common factor in several input-output transitions. One such transition leads from an input state that encodes the information that the sentence presented is 'Fa' to an output state that encodes the information that what was conveyed is that John is bald. Another leads from an input state encoding the information that 'Fb' was presented to an output state encoding the information that the message that Harry is bald was conveyed.

The pattern of derivational common factors in the deductive theory can be retained in an alternative theory that trades non-logical axioms for non-logical rules of inference. In a similar way, the pattern of causal common factors in a physical system can be retained in a system that trades task specific representations for task specific component processors.

The requirement that there should be a pattern of causal common factors that mirrors that pattern of derivational common factors created by the non-logical axioms of a deductive theory is clearly more permissive than the requirement of explicit representation of those axioms. But it is not a trivial requirement. There can, in principle, be two systems that achieve the same input-output function, where one system meets the requirement that there should be causal common factors corresponding to the rules of a task analysis, while the other system does not. The clearest case of failing to meet the requirement is provided by systems that perform a task on a case-by-case basis (by table look up).

Knowledge of rules in this intermediate sense requires that, where several input-output transitions conform to a rule, they should have a common causal explanation. That is, where several transitions exhibit the pattern described by a rule, they should involve a common component processor corresponding to that rule. (In essence, this is the notion of tacit knowledge to be found in Evans [1981]. For some refinements, see Davies [1987].)

If we define a syntactic property as a physical property that is both systematically related to semantics and a determinant of causal consequences (cf. Fodor [1987], pp.16-21), then it is possible to argue that knowledge of rules in the intermediate sense requires syntactic structure in input states. (See Davies [1989a], [1989b].) So the intermediate notion seems to capture something of what is involved in the classical approach to modelling cognitive processes.

The intermediate notion of knowledge of rules yields a corresponding notion of competence, located as "part and parcel of the processing". But, where knowledge of rules is embodied in component processors, there is not such an explanatory gap between having knowledge and using it. So, on the face of it, we shall expect a less interesting competence/performance distinction. Performance will fall short of competence just where the enabling conditions for the component processors fail to obtain.

This is not to say that competence is simply constituted by the input-output function that is achieved under ideal conditions. On the contrary, the intermediate notion of competence still plays an explanatory role. For the states of knowledge of the rules are implicated in the production of output states from given input states.

The intermediate notion of knowledge of rules makes some distinctions between connectionist networks. There can be two networks that each perform a task, where one meets the causal common factors condition, and the other does not. But the networks that meet the condition are particularly simple; in fact, the clearest cases involve local representation. Thus, for example, Figures 1 and 2 illustrate two very simple networks that perform the sentence interpretation task. The first does not meet the causal common factors condition: it performs the task by table look up. The second network does meet the condition.

However, this does not signal a successful conclusion to our search for connectionist competence. For, once we turn to networks with

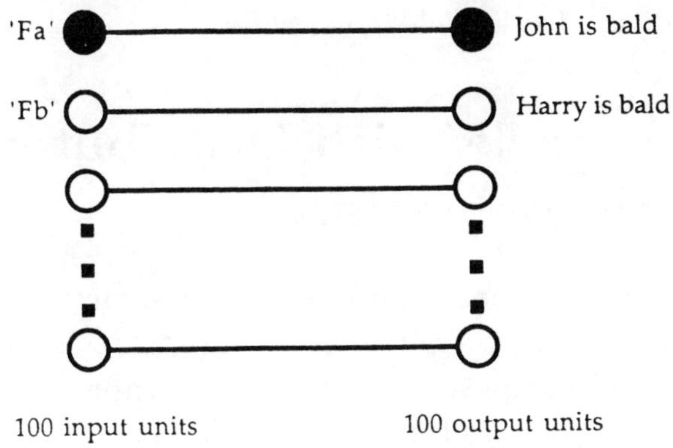


**Figure 1:** The first network presented with 'Fa'

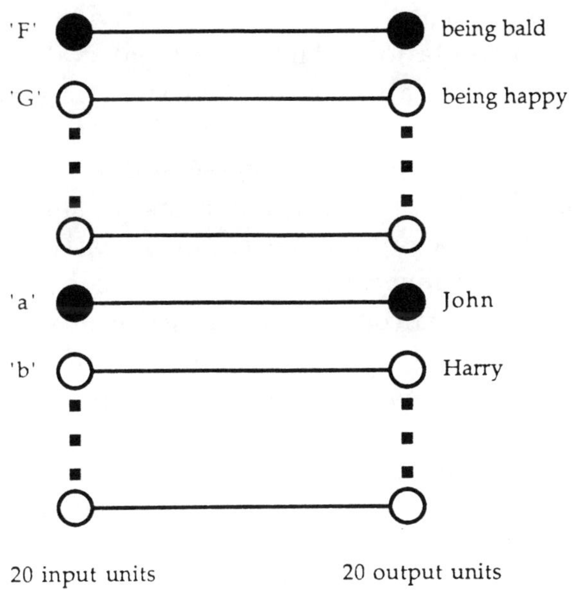


**Figure 2:** The second network presented with 'Fa'

distributed, microfeatural representation, there are quite general reasons for denying that connectionist networks meet the causal common factors condition for knowledge of rules.

## 5.4 Networks with Distributed Representation

Since knowledge of rules requires syntactic structure in input states, it suffices to argue that connectionist input states do not exhibit syntax. Input states are composed of activation at input units. The property of including activation at some particular unit or units is certainly a determinant of the causal consequences of an input state. But such properties are not systematically related to semantic properties cast in terms of the concepts used in the task analysis. For a primitive concept of the task analysis, there is no subpattern of activation which can be regarded as the tokening of a symbol expressing that concept.

Since representation is distributed, individual units are not the vehicles of representation. Thus, the input states of a system may be representing facts about coffee and wine, in cups and jugs, with or without sugar, but there is not an individual unit that represents the occurrence of coffee. The representation of coffee in a cup is not a matter of activation at a coffee unit and a cup unit. Rather, what represents coffee in a cup is a pattern of activation over many units.

Because of the "dimension shift" between concepts of the task analysis and microfeatural concepts, the constituent subpatterns that represent coffee in various contexts - coffee in a cup, coffee in a jug, coffee with sugar - are different (Smolensky [1988], p. 17):

> These constituent subpatterns representing coffee in varying contexts are activity vectors that are not identical, but possess a rich structure of commonalities and differences (a family resemblance, one might say).

(See also Smolensky [1987b], pp.145-8.) So there is no common distributed subpattern of activation that can be identified as a syntactic element meaning coffee. Relative to a semantic description in terms of coffee, wine, cups, jugs, and the like, the input states of a connectionist network with distributed representation and a "dimension shift" do not have syntactic properties.

We can also argue directly that, in general, there is not a causal common factor implicated in the various input-output transitions that conform to a rule uncovered by the task analysis.

A network may well achieve transitions from an input state meaning *coffee with sugar* to an output state meaning *warm drink with sugar*, from an input state meaning *coffee in a cup* to an output state meaning *warm drink in a cup*, and so on. What explains each of these transitions is a pattern of weights on connections. But it will not be true that there is a common set of weights on connections that is implicated in all the *coffee* to *warm drink* transitions.

This fact about typical connectionist networks does not, by itself, constitute an objection to connectionism. For, in general, it is an empirical question whether any given human cognitive process involves causal common factors.

It may be clear that a process is not simply a matter of table look up. But that is not nearly enough to show that the process involves knowledge of rules in our intermediate sense. It is clear enough that an ordinary English speaker's mastery of the past tense is not a matter of treating every verb as a completely separate case. But it is a further empirical question - calling for detailed evidence - whether, for example, mastery of the past tense of verbs that exhibit the regular pattern is a matter of a real commonality in causal explanations or only of family resemblance. It is this detailed evidence that is needed for an evaluation of a connectionist model of mastery of the past tense (such as the model of Rumelhart and McClelland [1986]).

The idea of causal common factors can, in principle, be used to distinguish between two systems with the same input-output relation. However, in real cases, it is overwhelmingly likely that a departure from the condition for knowledge of rules will show up somewhere in the input-output relation; particularly if we probe the system's operation by presenting novel inputs.

Let us briefly pursue the example of the past tense. If the transition from regular verbs to their past tenses is a matter of a causal common factor - a common component processor - then the input states for such verbs must have some common property to engage that common processor. (They must token a symbol meaning that the verb is regular.) If a new verb is presented then, provided that the input state has that same property (tokens that symbol), the verb will be awarded a past tense in just the same way as all other regular verbs.

The situation will be very different if there is only a family resemblance among the transitions for regular verbs; a family resemblance dictated by similarity of input states that are patterns of activation over units that respond to microfeatures of some kind. Then, if a new verb is presented, the transition to a past tense will be conditioned by the microfeatural similarity of the new verb to others. If the new verb is microfeaturally very different from other regular verbs, then it is likely to be awarded a past tense in a very different way.

In this case, we can see clearly why it is no objection to connectionism that networks do not exhibit knowledge of rules. For the empirical evidence might turn out to support the view that human performance of the past tense task is a matter of family resemblance, rather than knowledge of the rules uncovered by a classical linguistic analysis of the task. As it happens though, Pinker and Prince [1988] present evidence that human performance is not just a matter of family resemblance (see pp. 109-110, 123-25). It is this evidence, rather than any global preference for knowledge of rules, that provides the grounds for a negative evaluation of Rumelhart and McClelland's model.

## 5.5 Conformity to Rules

It is agreed on all sides that a distinctive feature of connectionism is that networks do not have knowledge of rules in the sense that requires explicit representation. We have just seen that there are reasons for denying to networks knowledge of rules even in the more permissive sense that requires only causal common factors, such as common component processors. So it must seem that the only notion of knowledge of rules that is applicable to connectionist networks is that according to which knowledge of a rule requires no more than behaviour that conforms to the rule.

This third notion of knowledge of rules provides for a corresponding

competence/
performance distinction. For it may be that the input-output behaviour of a system conforms to the rules only under idealisation. Under normal operating conditions the behaviour might conform to the rules only approximately.

This notion of competence is importantly different from the other two, since it plays only a descriptive, rather than a causal explanatory, role. If competence is constituted by performance under ideal conditions, then competence cannot be causally explanatory of performance. This notion can be applied straightforwardly to connectionist networks; but it is a very far cry from any conception that Chomsky may have intended.

Smolensky's [1988] account of the connectionist competence/ performance distinction is as follows (p.20):

> We have an inference system . . . whose performance is completely characterizable at the subconceptual level in terms of standard subsymbolic computation: massively parallel satisfaction of multiple soft constraints. . . . Just the same, the behaviour of the system can be analyzed at a higher level, and, under appropriate situations . . . , and under suitable processing idealizations . . . , the competence of the system can be described in utterly different computational terms.

(See also Clark's [1990] discussion of "Newtonian competence".)

The "processing idealizations" here include removal of time limitations. Under normal operating conditions, a pattern completing network that is given a part of some pattern might not complete it in the way required by the task analysis. But the competence of the network can still be characterised in terms of that task analysis, provided that the pattern would be correctly completed if the network were allowed unlimited time to settle into an output state. This is just what we expect, if competence is idealised performance.

The restriction to "appropriate situations" calls for separate comment. The appropriate situations are those in which the input state of the network poses a problem that falls properly within the task domain.

If an input state presents inadequate information, or an illicit combination of information, then there is no question of the production of an output state conforming, or failing to conform, to the rules of the task analysis. Yet it is a feature of networks that they produce "sensible" output states in these situations. So there is an aspect of a network's performance that outruns description in terms of conformity to rules.

On this conception, a task analysis can serve as a competence theory by providing an approximate description - both idealised and partial - of the input-output behaviour of a network that is successfully finding its way around the task domain.

It is a disputed question whether Smolensky's examples of sensible performance upon inadequate or inconsistent input states really distinguish connectionist networks from classical machines. (See Fodor and Pylyshyn [1988], pp.57-9.)

It is not at all clear, for example, that even a maximally explicit machine will, if given inconsistent input information "fall flat and deduce just anything at all" (Smolensky [1988], p.20). It is true, of course, that in classical logic any proposition at all can be derived from a contradictory premise. But the inferential processes in a maximally explicit machine will follow the lines of some proof procedure; and it will not in general be true that, from an inconsistent premise the proof procedure will yield "just anything at all".

But, however this may be, it is surely correct that networks with distributed microfeatural representation will often produce performance that outruns their competence.

Imagine, for example, that we have a typical distributed connectionist network that performs our sentence interpretation task. Instead of having input units dedicated to individual sentences (as in Figure 1), or to individual predicates and names (as in Figure 2), the network has input units that respond to visual features of the predicates and names. Consequently, the presence of the predicate 'F' in a presented sentence is represented by a pattern of activation over many input units. What is more, since 'F' is written differently in different contexts - that is, when paired with different names - the representations of 'F' in different sentences vary: there are commonalities and differences - a family resemblance. Suppose too that the output representations are distributed, so that the property of being bald, for example, is represented by a (context dependent) pattern of activation over units expressing semantic features of some kind.

If this network is presented with a completely new sentence that is made up of the predicate that is visually intermediate between 'F' and 'H', and a name that is visually intermediate between 'a' and 'c', say, then it will - we may suppose - produce a novel but interpretable output. It will be possible to explain why that output is produced given that input, in terms of the fundamental description of the network in terms of units and connections, activations and weights.

But the network's performance in assigning an interpretation to this novel sentence falls outside the scope of description in terms of conformity to the rules uncovered by the task analysis (the systematic theory of meaning). For those rules have no logical consequences for sentences other than the original one hundred.

Here then - on the path of conformity to rules rather than implicit rules - we have found a connectionist competence/performance distinction. We can leave it as an open question whether what we have before us is likely to "heal a deep and ancient rift in the science and philosophy of mind". But it is a disappointing feature of this account of competence that it moves the notion so far from its explanatory home base. Competence as conformity hardly honours its Chomskyan ancestry.

At the heart of the account is the idea that a task analysis can serve as a competence theory by providing an approximate description of a network's performance. We want to improve upon the disappointing feature, and move the notion of competence in an explanatory direction. So let us take the idea of approximation and combine it with the notion of implicit rules, instead of mere conformity to rules.

We had a general argument to show that networks - other than very simple localist networks - do not meet the causal common factors condition. But that argument does not rule out the possibility that, in some networks with distributed representation, the condition is approximately met.

## 5.6 Approximation and Explanation

Sometimes the processing in a network approximates the pattern of causal common factors that is required by the intermediate notion of knowledge of rules. Similarly, a network may approximate the conditions for syntactic structure in its input states: it may be approximately true that there is a common subpattern of input unit activation corresponding to a primitive concept of the task analysis.

Here we move from a claim about approximate description of performance to one about approximate explanation of performance. Knowledge of rules in the intermediate sense is a matter of the mechanisms and processes that produce the performance that conforms to the rule. To say that it is approximately true that a network has knowledge of a rule in the intermediate sense is to offer an approximately correct explanatory account of the performance that conforms to the rule.

So, suppose that there are approximate commonalities at the level of the formal description of the network - the description in terms of units and connections, activation and weights. Then, in our search for connectionist competence, either we can focus attention exclusively upon the network level of description, or else we can consider higher levels of description. If we consider higher levels of description of networks, then we face two possible cases.

On the one hand, it may be that there is no other vindicated level of description at which we do better than approximately meeting the causal common factors condition. Then, since what is approximately true at the network level of description is no better than approximately true at any other level, it is, in a word, false.

In this first case, the approximate characterisation - although false - need not be robbed of all theoretical interest. For the approximately correct explanations of the network's operation that are provided may well be crucial; their greater accessibility may be a worthwhile trade-off against their inaccuracy. Indeed, in many cases the alternative strictly correct explanations may be so long and complex as to be literally unintelligible to us. (See Smolensky [1987], p. 155.)

On the other hand, it may be that, while there are only approximate commonalities at the level of formal description of the network, there are real commonalities at some vindicated higher level of description.

Here, we do not have a relation of approximation between two accounts at a single level, but a relation of realisation or constitution between two levels. At the higher level of description, there is a device that has knowledge of the rules of the task analysis; a device whose input states have a syntax. At the lower level we have a connectionist realisation or constitution of that device.

(This second case is not a matter of "mere implementation" of a maximally explicit machine. In the device that we may find at a higher level of description of a typical connectionist network, the knowledge of rules is embodied in component processors rather than in explicit representations.)

In this case, it would be wrong to say that it is only approximately true that the network has knowledge of rules. For, if we consider the network under its higher level description, then it is true *simpliciter* that it has knowledge of rules. It is a mistake to suppose that the description pitched at the higher level is, in virtue of that fact, not *really* true. (See Block's [1989] discussion of "the Reductionist Cruncher".) If a description were rendered untrue simply by being at a higher level than some other description, then network descriptions themselves would not be true. Indeed, no descriptions would be true except those at the level of fundamental physics - if such there be.

It is this second case that promises us an explanatory notion of connectionist competence.

## 5.7 Higher Level Competence

Where a higher level description is vindicated, we can ascend from the network level of approximate, or blurred, commonalities to the higher level of real commonalities, for the purposes of psychological explanation in terms of knowledge of rules or competence. But doing so may place some of the network's performance outside the scope of that explanation.

It is often the fact that a network uses distributed representation - with overlapping representations of different concepts of the task analysis, and varying representations of the same concept in different contexts - that permits the network to generalise to novel cases: to extrapolate or interpolate. (See Hinton, McClelland and Rumelhart [1986], pp. 79-87 and Touretzky [1989] for discussion of some examples.)

So it is with our imagined example of the sentence interpreting network. In virtue of using distributed representation in terms of microfeatures, the network was able to assign an interpretation to a sentence visually intermediate between 'Fa' and 'Hc'. As we saw, this piece of performance falls outside the scope of description in terms of conformity to the rules of the systematic semantic theory.

Suppose now that we ascend from the network level of description, to a level at which the internal structure of, and the variation between, the distributed representations of individual predicates and names are ignored. We regard the network as realising or constituting a device with knowledge of the twenty semantic rules - one rule for each predicate and each name.

The fact that the device has knowledge of those rules explains, for example, why it produces an output that means that what was conveyed was that John is bald given a visual presentation of the sentence 'Fa'. But that competence does not explain the device's performance in assigning an interpretation to the novel sentence.

Consequently, if psychological explanation is pitched exclusively at the level of description where we find knowledge of rules and syntactically structured input states, then some of the performance of the device will fall outside the scope of that explanation. The device's generalisations beyond the one hundred sentences will be viewed, not as susceptible to psychological explanation in terms of knowledge about the task domain, but as the products of facts about realisation.

Quite generally, there will be aspects of a network's performance that fall outside the scope of a higher level competence explanation. On some occasions, these aspects will depart from human performance; in which case, it is so much the worse for the network's claim as a model of human cognition. But suppose that the network is a satisfactory match for human performance. Can we accept the network as a model, while insisting on pitching our competence explanations at a higher level of description?

There is no general answer to this question. Certainly, there can be no objection in principle to cognitive phenomena receiving non-cognitive explanations. For example, some co-occurrences of cognitive deficits are viewed as the products of facts about neuro-geography. Two deficits may appear in the same constellation of symptoms simply because their respective neural loci are both close to the site of an injury.

Also, there may well be phenomena which deserve a broadly psychological - rather than, say, physiological - explanation, but which do not demand an explanation in terms of knowledge about the task domain. These may include certain global properties of a system's performance.

Consider, for example, one of the phenomena covered by the term "graceful degradation": a connectionist network continues to perform tolerably well after internal damage such as destruction of some of the units or connections, or the addition of noise. This graceful degradation is not explained in terms of the competence that is revealed at a higher level of description of the network; knowledge of rules only provides explanations of the various input-output transitions that conform to the rules. But this is unsurprising since graceful degradation is not part of the task whose analysis yielded those rules.

So graceful degradation is a phenomenon that does not receive, but arguably does not demand, explanation in terms of what the system knows about the domain. The explanation of the phenomenon is, rather, in terms of the substructure in which the system realises its knowledge; namely, a distributed pattern of weights on connections.

It would be precipitate to refuse to allow that this is a broadly psychological explanation, and to insist instead that facts about connectionist substructure are, from the point of view of psychology, to be classed with facts about neuro-geography. After all, reaction times are the bread and butter of much cognitive psychology. And typical explanations of reaction time data are cast in terms, not of what the system knows, but of how that knowledge is stored, how it is accessed, what format is employed, and so on.

Sometimes, then, it is right that the explanation of one set of facts about a system's performance of a task should be of a quite different type from the explanation of another set. But there will surely also be cases where the partiality of higher level competence explanations is unacceptable.

Suppose that we accept that the interpolating performance of the sentence interpreter - its ability to deliver interpretations for sentences that are visually intermediate between sentences in the original class of one hundred - is part of its successful performance of the cognitive task. Then we shall expect that the interpolating performance is to be explained in terms of the same body of knowledge about the task domain as explains the performance with the original sentences. But this expectation is not met if we ascend to a level at which we ignore the distributed and blurred character of the network's representations, and find knowledge of just the twenty rules of the original task analysis.

In these circumstances, we need to redraw the boundaries of the cognitive task, and then to provide a new analysis, possibly employing different theoretical concepts. Knowledge of the new rules may be located at the same higher level of description as before, if the new analysis is cast in terms of the same theoretical primitives as before. It may be located at, or close to, the network level of description, if the new analysis is cast in terms of the microfeatures that are represented by individual input and output units. Or it may be located at some intermediate level. Analyses of the activity of the hidden units of a network, such as cluster analysis, may well contribute to the highly non-trivial task of constructing a new task analysis, to serve as a new competence theory.

This reverses the usual order of procedure, since it involves moving from model to theory, rather than from theory to model. It runs the danger of mistaking for a model what is really only a simulation. But in cases where the familiar top-down methodology has reached an impasse, or cases where the task domain is very ill understood, it may be a worthwhile strategy. (For some comments on the lack of distinctively connectionist task analyses, see Smolensky [1987], p. 156. For reversing the usual methodology, see Clark [1990].)

## 5.8 Varieties of Connectionist Competence

We began from the thought that different notions of knowledge of rules would provide for different notions of competence and of the competence/performance distinction.

The first divide that we confronted was between the explicit inaccessible rule view, and those views that locate competence as "part and parcel of the processing". Since the explicit inaccesible rule view has no application to connectionist networks, we followed the second alternative.

The next divide was between the implicit rule view, and the idea of competence as mere conformity. Since the implicit rule view is restricted to very simple (mainly localist) networks, we were temporarily driven all the way to the idea of knowledge of rules as mere conformity.

This idea is applied easily enough to networks. A task analysis with its rules and generalisations may serve as a theory of competence by providing an approximate - partial and idealised - description of the input-output behaviour of the network. But this notion of connectionist competence is unhappily far removed from a causal explanatory role.

Consequently, we took the idea of approximation and combined it with the implicit rule view instead of the conformity view, shifting attention from the behaviour of networks to the internal processes and mechanisms that explain the behaviour.

The implicit rule view (the intermediate notion of knowledge of rules) applies strictly only to very simple networks. But in other cases it may provide an approximately correct account of a network's internal architecture, and an approximately correct explanation of the network's rule fitting performance. If it is approximately true that a network has knowledge of rules, then a task analysis that articulates those rules can serve as an approximately correct theory of competence.

Approximation figures in two accounts of competence: in the account in terms of conformity to rules, and in the account in terms of implicit rules. But it figures in importantly different ways in those two accounts. In the first account, it may be really true that a network with distributed representation has a classical competence. For it may be really true that, if we both idealise and restrict attention to a part of the network's behaviour, then the behaviour conforms to the rules of a classical task analysis. In the second account, it is not strictly true that a network has a classical competence. For, however we idealise and restrict attention, it is not the case that the network's behaviour is produced by virtue of causal common factors. That is only an approximately correct account of the production of the behaviour. In the first account, approximation enters the story of what competence is. In the second account, approximation enters the story of the relation between competence and networks.

The first account provides a notion of competence that can apply to networks with distributed representation; but it is an unacceptably thin notion because it is merely descriptive. The second account provides an acceptably thick notion of competence, because it is causally explanatory; but it is a notion that only approximately applies to networks with distributed representation.

So long as we focus attention exclusively upon the proprietary network level of description, these two accounts seem to be the only contenders in our search for connectionist competence. But we can, instead, consider higher levels of description. This is the third divide that we faced.

If we do consider higher levels of description of networks, then we may sometimes be able to combine the attractive features of those two contending accounts.

Suppose that the acceptable thick notion of competence approximately applies at the network level of description. We then distinguish two possible cases. In one case, there is no higher level of description at which we do better than approximately meeting the conditions for implicit rules. But in the other case, we can ascend to a higher level of description of the network, where the causal common factor condition for the explanatory notion of competence really obtains.

In this second case, we have explanatory connectionist competence. But there will be aspects of the network's performance that fall outside the scope of the competence explanation. If that proves intolerable, then the way forward is to perform a new, more connectionist oriented, analysis of the cognitive task, and to credit a network with a new, less classical, competence.

## 5.9 An Alternative: Two-Tier Models

Since leaving aside the explicit rule view, we have been trying to locate competence as "part and parcel of the processing" in connectionist networks. An alternative approach is to locate competence in a system that is distinct from, though in some way harnessed to, the primary connectionist network. All the accounts so far have placed competence on-line; the alternative is to house competence off-line.

Very roughly indeed, the fundamental idea here is to employ a two-tier architecture. There is a connectionist network that performs a cognitive task, and there is a second system that exhibits classical competence by meeting the causal common factors condition relative to a classical analysis of that task. This second system might be a rule implicit machine - perhaps a localist network - or it might be maximally explicit.

The relationship between the two tiers might initially be conceived in either of two ways. On the one hand, the system with competence could be a back-up system for performing essentially the same informational transitions as the primary network. On the other hand, the competence exhibiting system could operate to assess the input-output performance of the primary network. That is, it could take as its input both the input and output of the primary network, and produce as output an assessment of the transition performed.

However, the first conception is not really adequate to the thought that the second system S embodies the competence of the primary network P. Of course, there can be dual route processing models in which one route is connectionist through and through, while the other route involves knowledge of rules. But, if we are to preserve anything at all of the intuitive idea of competence, then the operation of the system P without knowledge of rules must be in some way answerable to the operation of the system S with knowledge of rules.

So we should pursue the second conception: the system S assesses the performance of the connectionist network P. This does not exclude the possibility that S may also provide an alternative route for performing P's task. Indeed, one way for S to assess P's performance is by comparing it with S's own performance given the same input.

This second conception of a two-tier architecture is suggested by Norman ([1986], p.542):

> It is often the case that complex behaviour has to be monitored to see if things are proceeding well. This requires some sort of evaluation of the output of the system. But what mechanism does this evaluation? And what mecha-

> nism decides what the desired output is to be? I believe this requires an "overlooking" system, one that compares expectations with outcomes. This second, evaluative, system might very well also be a PDP structure, but it must be separate from the first, overlooking the first and operating upon it.

We can agree that an overseeing system could perform its evaluative task without meeting any causal common factors condition: it could operate by table look up, for example. But, what we are adding to Norman's suggestion is an idea about one way to develop a notion of off-line competence; namely, that knowledge of rules could be located in an overseeing system, rather than in the primary system.

## 5.10 Prospects for Two-Tier Models

Many questions about two-tier architectures deserve investigation. Here, we shall briefly take up just three topics: representational format, development, and normative aspects of thought and inference.

### *Representational format*

If the overseeing system S has knowledge of rules, then its input states must have syntactic structure. So, at the stage where the input information to the primary system P is made available to S, that information is encoded in a syntactic format. One possibility, then, is that the primary system itself has syntactically structured input states.

Thus, for example, in the case of the sentence interpretation task, the overseeing system with knowledge of the twenty semantic rules would use a syntactic format to encode the information that the presented sentence contains the predicate 'F' and the name 'a', say. The primary system could employ the same representational format. But, in order for either system to perform the task in the real world, these syntactically structured input states must be the product of a visual recognition system.

One very plausible kind of recognition device would be a little connectionist network whose input states use distributed representation over units tuned to visual microfeatures, and whose output states are syntactically structured, in readiness for the operation of S and P. But, if such a device is to be included in the total architecture, then it is clearly not obligatory that the primary system P should use the same syntactic format as S. An alternative possibility is that P should operate directly upon the distributed input states, and that the recognition device should be at the service only of S.

The two systems can, in short, employ different representational formats. If S is to oversee the performance of P, then the overall architecture must permit the reformatting of P's input information into S's preferred style. But that is a problem that would have to be solved in any case, just to put the operation of S in touch with the world. Similar remarks could be made about the format of output states.

The general picture of the relationship between S and P allows that S might, in fact, oversee the performance of two or more networks. Imagine, for example, that there are two sentence interpretation networks, one for visually presented sentences, and another for auditory presentations. Armed with both a visual and an auditory recognition device, S could oversee both networks. And reformatting the input information for S's use would be a step towards freeing the information from the constraints of a single sensory modality.

## *Development*

The idea of two-tier architectures need not be committed to any claim about developmental priority. The two broad possibilities are these. On the one hand, a symbol manipulating system with connectionist peripheral devices could be in place before the development of a network directly linking the microfeatural input and output states. On the other hand, a network might serve for dealing fluently enough with a task domain, while syntax and implicit or explicit rules arrive as a later construction.

The appeal of the first possibility is that it promises some help with the problem of supervised learning. The training of a network is apt to require thousands of trials, with the network's performance being monitored, errors being detected, and tiny alterations being made to the weights on connections (by back propagation of error, for example).

But this introduces a real implausibility when networks are offered as models for human cognitive processing. For, in many cases (of which language is perhaps the most obvious), human learning is fast, and depends mainly on positive evidence rather than error detection.

One way around this problem is to go nativist: to have the weights in a network largely endogenously determined. But the first possibility for developmental sequencing of S and P presents another option. The overseeing system can supervise the training of the primary network. Indeed, this is Norman's motivation for the suggestion of two-tier architectures.

The appeal of the second possibility is that it sits very happily with several intuitive ideas about cognitive development. One such idea is that connectionist networks seem well suited for sensori-motor coordination, while symbol manipulation machines may be just what is needed for conceptualised thought and inference. But sensori-motor coordination precedes conceptualised thought and inference, and arguably it is an important fact about the domain of thoughts that "the concepts exercised in the thoughts are learned by an organism in which the links between sensory input and behaviour are already established" (Evans [1982]. p.159; and cf. Peacocke [to appear b], Cussins [1990]).

Another idea that comes into play here is that cognitive development is partly a matter of the child coming to make better use of existing information by relaxing or transcending constraints. Thus, for example, a capacity which, at an early stage, can only be deployed within the confines of a particular behavioural routine may, at a later stage, be put at the service of other projects. These two stages are clearly reminiscent, respectively, of the modality specific network for sentence interpretation, and the system with knowledge of rules about the semantic properties of sentences in whatever modality they may be presented. (See Clark and Karmiloff-Smith [in preparation].)

Perhaps the way to reconcile the appeal of the two possibilities is to note that the training of a network requires a sample set of input-output pairs, but it does not require that the links between the members of the pairs should exhibit causal common factors. Indeed, for supervised learning a look up table would suffice. Training a network does not require knowledge of rules; it just requires knowledge of the right answers for the sample cases. So the need for a solution to the problem of supervised learning does not at all rule out the possibility that the competence embodying system S should be a later development than the primary connectionist network P.

## *Normative aspects of thought and inference*

Our commonsense scheme of propositional attitude description may appear to involve a commitment to causal common factors in inferential transitions. Here is a simple example of the key idea.

A thinker who has the thought that Bruce is a bachelor appreciates that from this thought it follows that Bruce is unmarried. Likewise, the thinker appreciates that from the thought that Nigel is a bachelor it follows that Nigel is unmarried. These are not two unrelated inferential dispositions. On the contrary, they have a common explanation in terms of knowledge what it is for an object to be a bachelor. This single state of knowledge figures as a causal common factor in both the inferential transitions that the thinker is disposed to make. (See Evans [1981], p.132; Davies [1990], [to appear].)

A very natural response to this is to say that while the commonsense scheme does indeed involve a commitment to causal common factors, it is not committed to causal common factors in unreflective inferential transitions. For there is a clear difference between appreciating what follows from a thought - appreciating what would be a good inference - and actually making inferences.

Thus, for example, on Peacocke's account ([1989], [to appear a]) mastery of a concept often involves finding certain inferences primitively compelling in virtue of their form. It is very natural to gloss "in virtue of their form" as meaning that there is a common causal explanation for the various inferences that exhibit a common form being found primitively compelling. But still, there is a difference between finding an inference compelling and making an inference.

The upshot is that the commonsense scheme's commitment properly concerns the assessment of inferences, rather than actual inferential transitions. What is required is a system that oversees the primary inferential device, and assesses the inferential transitions that are performed. It is this system that should exhibit knowledge of the rules uncovered by a theory of concept mastery. Many actual inferential transitions could proceed via a connectionist network whose input and output states are patterns of activation over units encoding features that are associated with various concepts.

These brief reflections may perhaps suggest some fair prospects for the idea of two-tier models which place competence off-line, but still firmly on the stage of psychological explanation.

We began with a puzzle. How can there be connectionist competence, when connectionist networks do not have knowledge of rules? For Chomsky introduced the notion of competence as equivalent to knowledge of rules. The solution was to distinguish the explicit inaccessible rule view from an account of rules as implicit - as "part and parcel of the processing itself". Indeed, before turning to two-tier models, we were able to develop a notion of higher level competence that is truly applicable to some distributed connectionist networks, and which retains the kind of explanatory role that Chomsky would insist upon.

In a recent discussion of the supposed analogy between the visual system and the language system, Chomsky [1988] remarks (pp.159-160):

> The visual system is unlike the language faculty in many crucial ways; it does not yield a system of knowledge, for example, but is strictly a processing system.

Perhaps this suggests one more prospect for two-tier models. By separating the system with knowledge of rules from the on-line processing system, these models promise a more truly Chomskyan notion of connectionist competence.

## References

Block, N., [1989]: "The Computer Model of the Mind", in Osherson, D., and Smith, E.E., (eds.), *An Invitation to Cognitive Science Volume 3: Reasoning*, MIT Press. Cambridge, Ma.

Clark, A., [1990]: "Connectionism, Competence and Explanation", *British Journal for the Philosophy of Science*, Vol. 41. Reprinted in Boden, M., (ed.), *The Philosophy of Artificial Intelligence*, Oxford University Press. Oxford.

Clark, A., and Karmiloff-Smith, A., [in preparation]: "The Cognizer's Innards"

Chomsky, N., [1965]: *Aspects of the Theory of Syntax*, MIT Press. Cambridge, Ma.

Chomsky, N., [1988]: *Language and Problems of Knowledge*, MIT Press. Cambridge, Ma.

Cussins, A., [1990]: "The Connectionist Construction of Concepts", in Boden, M., (ed.), *The Philosophy of Artificial Intelligence*, Oxford University Press. Oxford.

Davies, M., [1987]: "Tacit Knowledge and Semantic Theory: Can a Five Per Cent Difference Matter?" *Mind* vol. 96, 441-462.

Davies, M., [1989a]: "Connectionism, Modularity, and Tacit Knowledge", *British Journal for the Philosophy of Science*, vol. 40. 541-555.

Davies, M., [1989b]: "Knowledge of Rules in Connectionist Networks", to appear in a special issue of *Intellectica*, edited by Memmi D., and Visetti, Y.M.,

Davies, M., [1990]: "Thinking Persons and Cognitive Science", to appear in a special issue of *AI and Society*, edited by Clark, A., and Lutz, R.

Davies, M., [to appear]: "Aunty's Own Argument for the Language of Thought"

Evans, G., [1981] "Semantic Theory and Tacit Knowledge", in Holtzman, S., and Leich, C., (eds.), *Wittgenstein: To Follow a Rule*, Routledge and Kegan Paul. London. Reprinted in Evans, G., *Collected Papers*, 321-442, Oxford University Press. Oxford.

Evans, G., [1982]: *The Varieties of Reference*, Oxford University Press. Oxford.

Fodor, J., [1985]: "Fodor's Guide to Mental Representation", *Mind*, vol. 94, 77-100.

Fodor, J., [1987]: *Psychosemantics*, MIT Press. Cambridge, Ma.

Fodor, J., and Pylyshyn, Z., [1988]: "Connectionism and Cognitive Architecture: A Critical Analysis", *Cognition*, vol. 28, 3-71.

Hinton, G.E., McClelland, J.L., and Rumelhart, D.E., [1986]: "Distributed Representations", in Rumelhart, D.E., McClelland, J.L., and the PDP Research Group, *Parallel Distributed Processing: Explorations in the Microstructure of Cognition Volume 1: Foundations*, 77-109, MIT Press. Cambridge, Ma.

McClelland, J.L., Rumelhart, D.E., and Hinton, G.E., [1986]: "The Appeal of Parallel Distributed Processing", in Rumelhart, D.E., McClelland, J.L., and the PDP Research Group, *Parallel Distributed Processing: Explorations in the Microstructure of Cognition. Volume 1: Foundations*, 3-44, MIT Press. Cambridge, Ma.

Norman, D.A. [1986]: "Reflections on Cognition and Parallel Distributed Processing", in McClelland, J.L., Rumelhart D.E., and the PDP Research Group, *Parallel Distributed Processing: Explorations in the Microstructure of Cognition. Volume 2: Psychological and Biological Models*, 531-546, MIT Press. Cambridge, Ma.

Peacocke, C, [1989]: "Possession Conditions: A Focal Point for Theories of Concepts", *Mind and Language*, vol. 4, 51-56

Peacocke, C., [to appear a]: "Content and Norms in a Natural World", in Villaneuva E., and Valdivia L., (eds.), *Information-Theoretic Semantics and Epistemology*, Blackwell. Oxford.

Peacocke, C., [to appear b]: "Scenarios, Concepts and Perception"

Pinker, S., and Prince, A., [1988] "On Language and Connectionism: Analysis of a Parallel Distributed Processing Model of Language Acquisition", *Cognition* vol. 28, 73-193.

Rumelhart, D.E., and McClelland, J.L., [1986]: "On Learning the Past Tense of English Verbs", in McClelland, J.L., Rumelhart D.E., and the PDP Research Group, *Parallel Distributed Processing: Explorations in the Microstructure of Cognition. Volume 2: Psychological and Biological Models*, 216-271, MIT Press. Cambridge, Ma.

Smolensky, P., [1987]: "The Constituent Structure of Connectionist Mental States", *The Southern Journal of Philosophy,* vol. 26 Supplement, 137-161.

Smolensky, P., [1988]: "On the Proper Treatment of Connectionism", *Behavioral and Brain Sciences*, vol.11, 1-74.

Touretzky, D.S., [1989]: "Connectionism and Compositional Semantics", *Technical Report CMU-CS-89-147*, School of Computer Science, Carnegie Mellon University.

# Chapter 6

# Rule-Based Systems Ruled Out?

Tomáš Havránek
*Institute of Computer Science, Czechoslovak Academy of Sciences, Prague*

The aim of our research is **not** to convince users that they should use rule-based or MYCIN-like systems; these systems, even if primitive, seem to be still rather popular and seem not to give bad results (perhaps uncertainty processing is not really used). Our questions have been why this is the case and, provided that one wishes to use such systems where uncertainty measures are significant; what alternatives are open to one?

## 6.1 Rule-Based Systems with Numerical Uncertainties

Rule-based expert systems [Buchanan and Shortliffe, 1984] are still rather popular and widely used. They use knowledge bases consisting primarily of rules:

IF (antecedent) THEN (consequent) WITH WEIGHT (w).

In the sequel we shall write $A \Rightarrow C(w)$ only. Propositions that appear only on left sides of these rules are questions; they are to be evaluated in consultation with a user. Propositions that appear only on the right sides of the rules are potential goals of a consultation (under a given knowledge base). Other propositions are intermediate or inner and they are not visible to a user in an ordinary consultation run.

The weight $w$ of a rule is usually interpreted as follows:

> if the antecedent of the rule is known to be true then the degree of certainty $c$ that the rule gives to the consequent equals $w$.

Clearly, if the antecedent is known to be true only with some certainty $a$, then $c$ is given by a function $CTR$ of $a$ and $w$, and it is less than $w$.

In the common backward mode, a consultation system works as follows: it evaluates a goal by seeking all rules leading to that goal and trying to evaluate their antecedents; recursively all necessary propositions are evaluated, finally by asking questions.

What is the nature of weights, certainties or uncertainties, linked with the rules in a knowledge base? In the knowledge acquisition process, i.e. in constructing a knowledge base, there is a strong temptation to use some statistical knowledge for establishing weights, for example, point estimates of conditional probabilities $P(C \mid A)$ if they are available. This point of view is linked with the frequentists approach to probabilities. On the other hand, frequently, the weights are considered to be conditional subjective probabilities of a human expert.

This chapter is primarily a review of recent results obtained in the theory of rule-based systems. We are convinced that these results are relevant for critical evaluation of the uncertainty processing in rule-based expert systems both from subjectivist's and frequentist's point of view. For such an analysis it is necessary to bear in mind that rule-based systems are very appealing due to their computational feasibility and transparency; they are able to work with knowledge bases containing hundreds of propositions even if the construction of such a knowledge base is a complicated task. Moreover, practical experience with such systems is available, for our case with the EQUANT shell [Hájek and Hájková , 1984], [Hájek et al., 1989]. Investigations referred in this chapter can deepen our understanding of flaws of these systems.

In a consultation run of a rule-based expert system it is vital to be able to combine knowledge and degree of certainty, obtained for a proposition from two or more rules, i.e. from two or more information sources. This is done usually by a combining operation $\oplus$ that combines certainties $a$ and $b$ (from two sources) into the combined certainty $c = a \oplus b$ for the proposition in question. Such an operation is usually associative and commutative.

This approach is **extensional** since it pays no attention to possible relations between information sources, i.e. *mutatis mutandis*, an assumption concerning independence of sources in employed.

Hajek [1985] established an algebraic theory for such operations; this theory covers known combining operations such as those used by MYCIN or PROSPECTOR (see [Buchanan and Shortliffe, 1984]). If one considers (un)certainties as numbers from the open interval $(0, 1)$ with the neutral element $1/2$, the PROSPECTOR operation can be written as

$$a \oplus b \;=\; 1/(1 + (\frac{a}{1-a} * \frac{b}{1-b})^{-1}). \qquad (6.1)$$

This simply multiplies the odds $a/(1-a)$ and $b/(1-b)$.

In the algebraic theory it is assumed that the domain $D$ of certainties (not necessarily numbers!) is linearly ordered, has a greatest element, say **1** (certainly **yes**), a least element, say **0** (certainly **no**) and a neutral element, say **1/2**. The elements **0** and **1** are the extremal ones; the behavior of the operation $\oplus$ is to be defined as follows: for each $a \neq \mathbf{0}$, $\mathbf{1} \oplus a = \mathbf{1}$, and for each $a \neq \mathbf{1}$, $\mathbf{0} \oplus a = \mathbf{0}$. Result of $\mathbf{0} \oplus \mathbf{1}$ is left undefined. The operation $\oplus$ makes the non-extremal elements of $D$ to be an ordered Abelian group (OAG). The theory shows that all extensional rule-based systems work in a similar way and that in principle their work is independent of the choice of the combining operation. Clearly, the weights in the rules have to be in real applications chosen with respect to the operation actually used since a change in the combining operation can drastically change the results of a consultation run as concerns, for example, the ordering of goals according their resulting weights (cf. [Havránek and Soudský, 1988] and Section 6.2 here). On the other hand, it is clear from the theory, that problems linked with using statistically estimated certainties (rule weights) interpreted as estimated conditional probabilities in the frequentist's approach are inherent for all these systems, independently of the operations suggested for various shells.

To point out problems, we shall consider in the remainder of this section one-step systems, i.e. systems with rules leading directly from answers to goals. For more general systems the problems are considerably more complicated. To simplify matters, we shall make three restricting assumptions: first, we shall investigate, as we mentioned above, only knowledge patterns with no intermediate propositions. Thus propositions are divided into questions and goals; each rule has its antecedent, which is an elementary conjunction of questions, and the consequent, which is a goal. $QUES$ and $GOAL$ denote the set of all questions and goals respectively; $EC(QUES)$ is the set of all elementary conjunctions of questions, including the empty conjunction. Secondly, we shall consider only the PROSPECTOR combining operation $\oplus$ defined by (1) over the open interval $(0,1)$. Third, we shall allow the user to answer the questions only using 1, 0 and 1/2. This makes it possible to identify questionnaires (mappings $q : QUES - L(1, 1/2, 0)$) with elementary conjunctions of questions, e.g. if $q(1) = 1$, $q(2) = 0$ and $q(i) = 1/2$ for other questions then $q$ is identified with $1\&\neg 2$. Note that there is no reason to exclude the empty antecedent; $\Rightarrow C(w)$ means that the *a priori* weight of $C$ is $w$. Consider now rules $E_1 \Rightarrow H$ and $E_2 \Rightarrow H$ where $H$ is a goal and $E_1$ and $E_2$ are antecedents having no common question. The PROSPECTOR combining operation was derived under the assumption of conditional independence of $E_1$ and $E_2$ conditioned by the goal $H$. If there is no misunderstanding possible we shall denote by $A$ the random event "A is true" etc. Then the conditional independence assumption means that $P(E_1 \wedge E_2 \mid H) = P(E_1 \mid H)P(E_2 \mid H)$ and $P(E_1 \wedge E_2 \mid \neg H) = P(E_1 \mid \neg H)P(E_2 \mid \neg H)$. In [Hájek and Havránek, 1987] some unpleasant consequences of the independence assumption are studied (the main theorem was published earlier by [Johnson, 1986]). We are using here the letter $E_i$ for antecedent, of rules which are elementary conjunctions of questions, to link antecedents with items of **evidence** and the letter $H_j$ for a goal, since one can interpret goals as **hypotheses**.

Let us now investigate simple knowledge bases consisting of disjoint items of evidence $E_i (i = 1, ..., m)$, goals $H_j (j = 1, ..., n)$ and rules $E_i \Rightarrow H_j (i = 1, ..., m, j = 1, ..., n)$ together with numbers $a_i$, $c_j$, $r_{ij}$, $\in (0, 1)$ interpreted as prior probability of $E_i$, of $H_j$ and of the rule $E_i \Rightarrow H_j$ respectively. Thus we call an **(m,n)-knowledge base** any system of numbers $a_i$ , $b_j$ and $r_{i,j}$ from the open interval $(0, 1)$. A probability distribution $P$ on $E_1, ..., E_m$, $H_1, ..., H_n$ realizes a knowledge base if $P(E_i) = a_i$, $P(H_j) = c_j$, $P(H_j \mid E_i) = r_{ij}$ for $i = 1, ..., m, j = 1, ..., n$.

The following fact can be easily proven: If a knowledge base is realizable then

$$c_j/a_i \geq r_{i,j} \geq (c_j + a_i - 1)/a_i \tag{6.2}$$

for all $i, j$. A knowledge base satisfying the conditions (6.2) is called **sound**. Let $1E_i$ be $E_i$ and $0E_i$ be $\neg E_i$ ; if $\mathbf{u} = (u_1, ..., u_m)$ and $E = (E_1, ..., E_m)$ then $\mathbf{uE}$ denotes the elementary conjunction $u_1 E_1 \wedge ... \wedge u_m E_m$.

A probability $P$ on $\mathbf{E}$, $\mathbf{H}$ is PROSPECTOR-like if, for each $\mathbf{u}$ and each $j$, $P(\mathbf{uE} \mid H_j) = \prod_{i=1}^m P(u_i E_i \mid H_j)$ and $P(\mathbf{uE} \mid \neg H_j) = \prod_{i=1}^m P(u_i E_i \mid \neg H_j)$. $P$ is exhaustive if $\sum P(H_j) = 1$ and for each $i$, $\sum P(H_j \mid E_i) = 1$; $P$ is exclusive if $P(H_j \wedge H_k) = 0$ for any $j \neq k$.

Now we can express the main theorem mentioned above:

> Let $\mathbf{B} =< a_i, c_j, r_{i,j} >$ be a sound $(m, n)$-knowledge base with $m \geq 2$ and $n \geq 3$, let $\sum_j c_j = 1$ and, for each $i$, $\sum_j r_{i,j} = 1$. There is a PROSPECTOR-like exclusive and exhaustive probability realizing $\mathbf{B}$ iff for each $j = 1, ..., n$ there is an $i_0 \in \{1, ..., m\}$ such that for each $i = 1, ..., m$ except possibly $i_0$ , $E_i$ and $H_j$ are stochastically independent.

To summarize, under very simple and reasonable assumptions, each goal depends on at most one item of evidence!

In knowledge bases, where disjoint items of evidence, and antecedents, are not the only ones usually used, this is a source of further problems. Consider an example from [Hájek, 1984]: Imagine a miniature knowledge base consisting of three rules $E_1 \Rightarrow H(w_1)$, $E_2 \Rightarrow H(w_2)$ and $E_1 \wedge E_2 \Rightarrow H(w_3)$. If a user in a consultation run is certain completely that $E_1$ is true and he does not know anything about $E_2$ then the global weight $W(H \mid E_1)$ of $H$ under this information is $w_1$ ; only the first rule is operative and its contribution $c_1$ is just $w_1$. Similarly if 1 is replaced by 2. Now assume that both $E_1$ and $E_2$ are known to be true. Then the global weight under this information is $W(H \mid E_1 \wedge E_2) = w_1 \oplus w_2 \oplus w_3$. If all $w_1$, $w_2$ , $w_3$ are greater than 1/2 (unknown) then $W(H \mid E_1 \wedge E_2) > max\{w_1, w_2, w_3\}$. Now consider that $w_1, w_2, w_3$ are estimates of the conditional probabilities $P(H \mid E_1)$, $P(H \mid E_2)$ and $P(H \mid E_1 \wedge E_2)$ from a given data set. Then clearly under the information that $E_1$ and $E_2$ are true, $W(H \mid E_1 \wedge E_2) = w_3$ is to be used! Using the combining operation $\oplus$ this can be done, but the weight of the third rule is to be changed to $E_1 \wedge E_2 \Rightarrow H(w_3 \ominus w_2 \ominus w_1)$; the direct relation between weights and estimates is lost. The same remains true if weights are interpreted as subjective probabilities, the weight of the rule $E_1 \wedge E_2 \Rightarrow H$ is no more the expert's subjective probability of H under condition that $E_1$ and $E_2$ is true. In general, weights in complex rules are to be recomputed using weights of the simple ones.

As proved by [Hájek 1984] this approach is realizable for one-step knowledge bases, but intuitively expected properties of weights are lost (cf. Section 3). This approach is implemented in expert system shell ESOD of Ivánek and Stejskal [1988]. On the other hand, in EPROB Havránek and Kouba [1987], suggested that only the most complex rules are operative. Both these shells are proposed for the case when rules and weights are generated on the basis of a data set containing evaluation of propositions on various cases.

## 6.2 Uncertainty of Uncertainty

Following the frequentist's approach one easily recognizes that point estimates of weights (probabilities) are of variable quality depending, for example, on the range of data sets from which they are computed. But similar considerations can be made for the subjectivist's approach too; uncertainties, certainties, weights are again established with variable accuracy and reliability.

These considerations lead in general to substituting for a single numerical value, pairs of values. We now focus our attention on the

Dempster-Shafer theory (see e.g. [Dempster, 1967, 1968] and [Shafer, 1976]) as an interesting generalization of the probability theory repeatedly discussed in the context of expert systems. As noted by Valdés [1987] and Hájek and Valdés [1987a,b], Dempster semigroup $D$ operates on pairs of reals $< a, b >$ such that $0 \leq a, b \leq 1$ and $a + b \leq 1$ (excluding $< 1, 0 >$ and $< 0, 1 >$). The combining operation is:

$$< a, b > \oplus < c, d >= \left(1 - \frac{(1-a)(1-c)}{1-(ad+bc)}, 1 - \frac{(1-b)(1-d)}{1-(ad+bc)}\right) \quad (6.3)$$

This operation was suggested under the assumption that information from independent information sources is combined. Let us denote $a' = 1 - a$, $b' = 1 - b$. There is a strong temptation to interpret $[a, b']$ as a closed confidence interval for uncertainty linked with the actually evaluated proposition; the combining function gives an interval arithmetic on such intervals.

There is a natural linear order on $D$, making it to be an ordered semigroup. The semigroup $D$ contains an ordered subgroup $G \subseteq D$ isomorphic to the additive group of reals and which is an injective homomorphic image of $D$ [Valdés, 1987]. There are two distinguished elements in $D$, namely $< 0, 0 >= [0, 1]$ and $< 1/2, 1/2 >$. The first one can be interpreted as complete ignorance (we know nothing about the proposition in question), the second one as complete uncertainty (we know that proposition and its negation are equally probable). These elements are the only idempotent elements in $< D, \oplus >$.

Following [Valdés 1987] and [Hájek and Valdés 1987a,b], consider now the mapping $h$ on $D$ defined as $h(< a, b >) =< a, b > \oplus < 1/2, 1/2 >$. This algebraic definition gives the function $h(< a, b >) =< b'/(a' + b'), a'/(a' + b') >$; in the interval interpretation we obtain one point interval containing the point $c = b'/(a' + b')$ that can be taken as a **center** of the interval $[a, b']$.

The mapping $c : D \Longrightarrow \varphi(0, 1)$ is an injective homomorphism to an ordered Abelian group (**OAG**) with the Dempster operation (restricted to first components) and the natural linear ordering of reals; the neutral element being $1/2$.

As proved by [Valdés 1987], this group $G$ is isomorphic to the PROSPECTOR OAG in $(-1, 1)$ by $i(x) = 2x - 1$ and hence by [Hájek

1985] to the additive group of reals (see Section 1). The natural ordering on the Dempster semigroup is defined using centers: $< a, b > \leq_D < a_2, b_2 >$ if $c_1 < c_2$ or $(c_1 = c_2$ and $a_1 \leq a_2)$ .

[Valdés 1987] interprets $c$ as the certainty contained in the Dempster pair $< a, b >$. He defines another mapping expressing the **vagueness** of $< a, b >$. This mapping is defined algebraically in a simple way $f(< a, b >) = < a, b > \oplus < b, a >$. The $f$-image of $< a, b >$ is a pair of two identical elements, hence $f$ can be considered as a homomorphism of $D$ into the interval $[0, 1/2]$. It is important that for $< a, b > \in D$, $a = b$ iff $f(< a, b >) = 1/2$. The interval $[0, 1/2]$ with the operation $\oplus_f$ induced by $f$ forms a semigroup $S$ (a positive cone of an OAG plus a maximal element 1/2). This semigroup is isomorphic with the nonnegative part of the additive group a reals. The analytic expression for $f$ is not transparent: $f(< a, b >) = (a + b - ab - (a^2 + b^2))/(1 - (a^2 + b^2))$. Finally, note that $< h, f >$ forms an injective homomorphism of $D$ into $GXS$ (order preserved).

These structural properties of $D$ may be used [Valdés, 1987], [Hájek and Valdés, 1987a,b] to define the class of **dempsteroids** - special ordered semigroups (models of a simple first order theory) such that $D$ is just a particular (standard) dempsteroid. This is a natural generalization of the Hájek's theory of simple certainties to the "interval" case. Each dempsteroid can be thought of as just a refinement of an OAG $G$ that results by replacing each $g \in G$ by infinitely many pairs $< g, v >$, where $g$ is the **certainty** and $v$ is the **vagueness**. Structural properties of finitely generated dempsteroids are well established.

How is this approach realized in the expert system shell EQUANT-PC [Hájek et al., 1989]? A knowledge base consists of propositions (numbered by natural numbers $1, 2, ..., N$) and rules. Each rule has the form $A \Rightarrow C(w)$ where $A$ is an elementary conjunction of some propositions, $C$ is a consequent being a proposition not occurring in $A$ and $w$ is a **designated** value - an integer such that $-10 \leq w \leq 10$. Here $-10$ means 'certainly no', 10 means 'certainly yes' and integers $-9, ..., 9$ are intermediate **comparative** degrees of belief. The system of rules must be loop-free. During a consultation the user is asked to ascribe his degree of belief to some propositions (questions); he may answer by a designated value or, more generally, by an interval, i.e. by a pair of designated values $< d_1, d_2 >$ such that $d_1 \leq d_2$ ($d$ may be identified with pair $< d, d >$). By answering questions the user defines a questionnaire, i.e. a partial mapping of propositions into designated values (a mapping from propositions to intervals).

A **realization** of designated values is a one-to-one monotone mapping $r$ of designated values into the closed interval $[-1, 1]$ of reals such that $r(10) = 1$ and $r(-w) = -r(w)$. The interval is endowed with the operation $\oplus$ defined as follows: $x \oplus y = (x + y)/(1 + xy)$. The interval with this operation, the unary operation $-x$ and the usual linear ordering $\oplus$ of reals is the **standard (PROSPECTOR) combining structure**. The set $D = \{< a, b >; a, b \in [-1, 1] \text{ and } a \leq b\}$ is endowed with the operation $\oplus$ induced from the operation (3) by the mapping mentioned below. Further we can consider the induced centering mapping $h$ and define $< a, b > \leq < c, d >$ if $h(< a, b >) < h(< c, d >)$ or $(h(< a, b >) = h(< c, d >)$ and $a \leq c)$. The set $D$ with the operations $\oplus$, '-' and the ordering $\leq$ just defined is the **standard Dempster combining structure** (we transform pairs $< x, y >$ such that $0 \leq x, y \leq 1$ and $x + y \leq 1$ into intervals $(a, b)$, where $a = 2x - 1, b = 1 - 2y$).

The answer of the user plus the realization of designated values determine for each proposition $i$ which has been a goal of the consultation a unique element of $D$ (which belongs to $P = [-1, 1]$ if the user does not use intervals) called the **numerical weight** of $i$. The computation may be done by the usual backward chaining method, using $\oplus$ for joining contributions of rules with the same consequent. The comparative result of the consultation is the quasi ordering of goals according to their numerical weights. (This can be refined by telling for each goal that its weight lies strictly between realizations of the successive designated values $w, w'$ or that it coincides with the realization of $w$.)

This approach is general in the following sense (here we disregard intervals for simplicity): One can use realizations of designated values in each combining structure $G$ that results from an ordered Abelian group by adding a top element **1** and a bottom element **0**. Call $< G, r >$ and $< G', r' >$ **equivalent** for the knowledge base **B** if for each questionnaire $q$, the comparative result of the run given by $< \mathbf{B}, G, r, q >$ is the same as that given by $< \mathbf{B}, G', r', q >$. This is to say that if the user is told the comparative results he is unable to distinguish whether the inference machine computes with $< G, r >$ or with $< G', r' >$. Now the generality is based on the following theorem:

Let $\mathbf{B}$ be given. For each realization $r$ of designated values in a combining structure $G$ there is a realization $r'$ in $P = [-1, 1]$ with the Prospector $\oplus$ such that $< G, r >$ and $< P, r' >$ are equivalent for $\mathbf{B}$.

Hence, in the comparative sense, all possible combining structures (operations) are given by various realizations in the standard Prospector one.

For experiments a knowledge base for cluster analysis was used [Havránek and Soudský, 1988], [Hájek et al., 1989]. In a run, one can consider 10 goals from the total 48 potential ones, denoted by numbers 1 to 6, 9, 11, 17 and 26. The user can choose, for example, the Prospector equidistant realization of the designated values and four additional random realizations. After answering questions, the following overview can be obtained:

| realization | goals |
|---|---|
| 1 | 9 >17 = 3 = 1 = 2 >11 >26 > 4 = 5 = 6 |
| 2 | 3 = 1 = 2 =17 > 9 >11 > 4 = 5 = 6 >26 |
| 3 | 9 > 3 = 1 = 2 =17 >11 >26 = 4 = 5 = 6 |
| 4 | 9 > 3 = 1 = 2 >11 >26 =17 > 4 = 5 = 6 |
| 5 | 3 = 1 = 2 > 4 = 5 = 6 > 9 >17 >11 >26 |

For the ordering of goals, the ordering defined above was used. The user can see that for all realizations, e.g $9 > 11$ and $9 > 26$ etc., but on the other hand the results are dependent on realizations of designated values (i.e. on the choice of combining operation).

Even if a comparative approach in the presentation of the results of consultation runs is stressed, and if an evaluation of goals in terms of uncertainties is obtained, then it is necessary to order them according to the corresponding numerical degrees of belief. This is the fundamental paradigm concerning the presentation of results to users usually employed in constructing expert system shells. In our case, if $h$ is the centering mapping and $f$ the vagueness mapping, then the lexicographic quasi-ordering of intervals is used. First they are compared with respect to centers and only if centers coincide, then the vagueness is taken into account ($[a, b'] <_D [c, d']$ if $f(< a, b >) > f(< c, d >)$. The underlying idea is that the resulting order must be linear, i.e. each two goals have to be comparable (let us omit here the theoretical algebraic reasons). As a result we obtain that $[a, b'] <_D [c, d']$ (or goal $g_1 <$ goal $g_2$) even in the case when

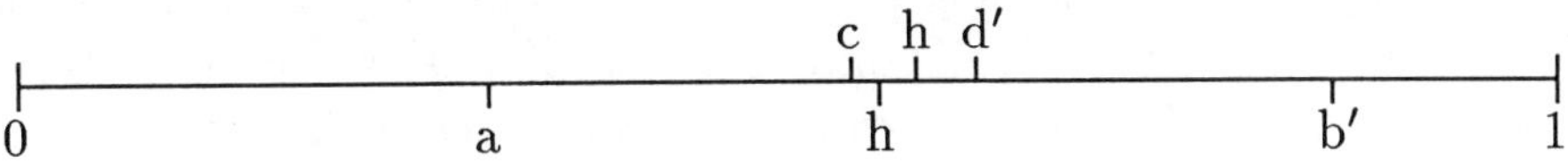


If the only knowledge is that the "probability" of $g_1$ is somewhere in $[a, b']$ and the "probability" of $g_2$ is somewhere in $[c, d']$, then the decision that $g_1 < g_2$ seems to be, without any additional assumptions, rather risky. The possibility of an error can have the chance close to 1/2.

By interval uncertainties only a partial quasi-ordering be reliably generated: $g_1 \leq_p g_2$ if $a \leq c$ and $b' \leq d'$. For each pair of intervals $[a, b']$ and $[c, d']$ there is an infimum $[a, b'] \wedge [c, d'] = [min(a, c), min(b', d')]$ and a supremum $[a, b'] \vee [c, d'] = [max(a, c), max(b', d')]$; a lattice is formed. Clearly, $<_p$ implies $<_D$. From the naive point of view, $\wedge$ is a candidate to be used in the contribution function CTR for firing rules (if $< a, b >$ is the uncertainty of the antecedent, $< v, w >$ is the weight of the rule, then the uncertainty of the consequent can be defined by $[a, b'] \wedge [v, w']$).

What is substantial here is that as a result of numerical manipulations with uncertainties only a (natural) **partial ordering** of goals is obtained. [Valdés 1989] considers partial ordering of propositions in a way similar to multicriterial decision making - his approach is based on vector uncertainties.

It is perhaps more substantial, and for knowledge bases representing human expert knowledge more natural, to **start with partial ordering of propositions** (goals) under some evidence established by a human expert. This can be closer to human intelligence than the usual subjective comparative probability approach demanding linear quasi-orderings perhaps still with a numerical representation in the background.

Another possible way is to abandon rule-based systems altogether and work with probabilistically based systems (e.g. [Perez and Jiroušek,

1985], [Spiegelhalter, 1986], [Lauritzen and Spiegelhalter, 1987]), although with possibly considerably greater computational complexity and less transparency for both experts and users. It is worth mentioning that even here some independence assumptions must be employed or some approximations must be used in implementing theoretical concepts. From the statistical point of view, if one wishes to estimate underlying probabilities (uncertainties) from data, these systems have clear advantages. Clearly, such an estimation process frequently demands work with contingency tables of high dimension. The problem of combining the data generated with human expert knowledge remains, since such cases can result in obtaining non-consistent probability distributions.

In both cases, and this is the lesson of Section 2, it is necessary to take into account the quality of the estimates - these estimates are to be used for decisions. In simple rule based systems here is a way of using "interval" estimates. An analogous way is to be applied to probabilistic systems too, but it can be considerably more complicated (cf. [Spiegelhalter, 1986b]). In both cases in the knowledge acquisition process based on statistical data, problems of simultaneous inference and model choice (see e.g. [Edwards and Havránek, 1985, 1987]) are to be solved. The later is linked with some "free" use of statistics for suggesting relations or rules based on data for further evaluation by human experts with respect to their role in a knowledge base.

## 6.3 What can MYCIN-like Systems be Good for from the Probabilistic Point of View?

Clearly, 'probabilistic' should be read here more from the frequentist's than from the subjectivist's point of view. To avoid prejudiced and/or simple minded answers [Hájek 1988,1989] elaborated a precise formal apparatus and formulated precise mathematical questions. At present, only rather particular answers are known, but it may not be difficult to find more complete answers.

We shall use restrictive assumptions from the end of Section 1. If

$\Theta$ is a knowledge pattern, i.e. a set of rules, and $\kappa$ a weighting, i.e. a mapping from $\Theta$ to $[0,1]$, then we write $\mathbf{B} = <\Theta, \kappa>$ for the knowledge base. For each goal $H$ and each elementary conjunction $E$ of some questions we have the **global weight** of $H$ given $E$ defined as follows: $W_{\mathbf{B}}(H \mid E) = \bigoplus_{E' \subseteq E, E' \in \Theta} \kappa(E' \Rightarrow H)$; note that the present setting makes the combining functions CTR (contribution of a rule), CONJ (evaluating of a conjunction) and NEG (evaluating of negation) superfluous.

Assume now $QUES$ and $GOAL$ given. A **belief system** is a mapping $\beta$ of a subset of $EC(QUES) \times GOAL$ into $[0,1]$. If $E \in EC(QUES)$, $H \in GOAL$ and $\beta$ is defined for the pair $<E, H>$ we denote the value by $\beta(H \mid E)$ resembling conditional probability. A belief system $\beta$ is total if $dom(\beta)$ is the whole of $EC(QUES) \times GOAL$; otherwise it is partial. A belief system could correspond to subjective views of a human expert concerning his/her belief in each $H$ conditioned by some $E$; on the other hand one can imagine a belief system as a result of estimation of conditional probabilities of $H$ under $E$. Clearly, in both cases the belief system could be only partial.

Two belief systems $\beta$, $\beta'$ are **numerically compatible** if for each $<E, H> \in dom(\beta) \cap dom(\beta')$ we have $\beta(H \mid E) = \beta'(H \mid E)$. They are **comparatively compatible** if for each $<E, H>$, $<E', H'> \in dom(\beta) \cap dom(\beta')$ we have $\beta(H \mid E) \leq \beta(H' \mid E')$ iff $\beta'(H \mid E) \leq \beta'(H' \mid E')$.

Let now $\mathbf{B} = <\Theta, \kappa>$ be a knowledge base corresponding to the usual representation of a human expert knowledge in rule-based systems. The belief system **generated** by $\mathbf{B}$ is the above $W_{\mathbf{B}}$. What is counterintuitive is the fact that if $E \Rightarrow H(w)$ belongs to $\mathbf{B}$ (i.e. the rule $E \Rightarrow H$ is in $\Theta$ and $\kappa(E \Rightarrow H) = w$) and $\beta = W_{\mathbf{B}}$ is the belief system **generated** by $\mathbf{B}$ then we need not have $\beta(H \mid E) = w$; see the example in Section 1. The weight of the rule is understood as the modification of the joint effect of simpler rules leading to $H$. Let $\beta'$ be an arbitrary belief system; $\beta'$ is numerically (comparatively) compatible with $\mathbf{B}$ iff $\beta'$ is numerically (comparatively) compatible with $W_{\mathbf{B}}$.

In a probabilistic approach one should consider, at least theoretically, a joint probability distribution over the elementary propositions. A **joint distribution** on $QUES \cup GOAL$ is a mapping $D$ associating with each elementary conjunction of all elements of $QUES \cup GOAL$

a real from $[0,1]$ such that $\sum_{K \in dom(D)} D(K) = 1$. The probability $P_D(K)$ of any elementary conjunction of elements of $QUES \cup GOAL$ is $P_D(K) = \sum_{K' \in dom(D) \& K \subset K'} D(K')$. The conditional probability $P_D(H \mid E)$ for $E \in EC(QUES)$ and $H \in GOAL$ is defined by $P_D(H \mid E) = P_D(H \wedge E)/P_D(E)$ for $P_D(E) > 0$. The mapping $\beta(H \mid E) = P_D(H \mid E)$ is a total belief system generated by the distribution $D$. A belief system $\beta$ is **numerically (comparatively) probabilistically sound** if there is a distribution $D$ such that $\beta$ is numerically (comparatively) compatible with $D$. A belief system $\beta$ is **weakly sound** if for each $E$, $H$ such that $\beta(H \mid E)$ is 0 or 1 and each $E' \supseteq E$ we have $\beta(H \mid E) = \beta(H \mid E')$ if defined.

Now we are able to formulate problems. Our starting point is the following theorem: If $\beta$ is weakly sound then there is a knowledge base $\mathbf{B} = <\Theta, \kappa>$ such that $dom(\beta) = dom(\kappa)$ and $\beta$ is numerically compatible with $\mathbf{B}$.

Note that if $\beta(H \mid E)$ is defined then it is enough to put $\kappa(E \Rightarrow H) = \bigoplus_{E' \subseteq E, \beta(H|E') \text{ defined}} (-)^{|E-E'|}\beta(H \mid E')$, where $(-)^{2n-1}p = 1 - p$ and $(-)^{2n}p = p$ (cf. the example in Section 1). Note that each numerically probabilistically sound system is weakly sound.

As a corollary we obtain: For each numerically (comparatively) probabilistically sound belief system $\beta$ there is a knowledge base $\mathbf{B} = <\Theta, \kappa>$ such that $dom(\beta) = dom(\kappa)$ and $\beta$ is numerically (comparatively) compatible with $\mathbf{B}$.

The condition on domains implies that the cardinality of the set of rules is the same as the cardinality of $dom(\beta)$. This looks nice **but** the problem is whether the total belief system $W_{\mathbf{B}}$ generated by $\mathbf{B}$ is probabilistically sound. Clearly, since $\beta$ is probabilistically sound we may first find a distribution $D$ such that $\beta$ is compatible with the belief system $P_D$ and then apply the corollary to $P_D$; thus we have a new corollary: Each numerically (comparatively) probabilistically sound belief system $\beta$ is numerically (comparatively) compatible with a total belief system $\beta'$ which is generated by a knowledge base $\mathbf{B}$ and is numerically (comparatively) probabilistically sound.

This again looks nice but now we have no control on the cardinality of the set of rules in $\mathbf{B}$. Thus we are led to the following two problems that appear to be of great importance for the correct evaluation of MYCIN-like or rule-based systems from the probabilistic point of view:

**Problem A**: Let $\beta$ be a belief system with rational values. How to decide simply whether $\beta$ is comparatively probabilistically sound? (Note that the question is decidable thanks to Tarski's result on the decidability of elementary algebra.)

**Problem B**: Assume that $\beta$ is comparatively probabilistically sound. How to find a possibly simple knowledge base **B** such that $\beta$ is comparatively compatible with **B** and the belief system $W$ generated by **B** is comparatively probabilistically sound?

Problem A is also relevant for the purely probabilistic approach to rule-based systems. Note that at the first glance it is possible that the computational complexity of either problem is high.

Some additional information concerning the problem A can be found in [Hájek, 1988]. Important results concerning the problem B are presented in [Hájek, 1988, 1989]. These results are based on graphic modeling of dependence structure in the joint probability distribution over propositions (see [Lauritzen and Spiegelhalter, 1988]). These results are interesting but up to now unsatisfactory since, firstly, they concern numerical values, not orderings, and, secondly, they concern only one-step systems. Satisfactory answers to our problems remain to be found.

## References

Buchannan, B. and Shortliffe, E.H., (eds.), [1984]: *Rule-based expert systems*, Addison-Wesley.

Dempster, A.P., [1967]: "Upper and lower probability inferences", *Ann. of Statistics 38*, 325-339.

Dempster, A.P., [1968]: "A generalization of Bayesian inference", *J. Royal Stat. Soc. B 30*, 208-247.

Edwards, D., Havránek, T., [1985]: "A fast procedure for model search in multidimensional contingency tables", *Biometrika* 72, 339-351.

Edwards, D., Havránek, T., [1987]: "A fast model selection procedure for large families of models", *J. Amer. Statist. Assoc. 82*, 205-213.

Hájek, P., [1984]: "Combining functions for certainty degrees in consulting systems - a remark", I. Plander (ed.), *Artificial Intelligence and Information Control Systems of Robots*, North-Holland, 163-166.

Hájek, P., [1985]: "Combining functions for certainty degrees in consulting systems", *Int. J. Man-Machine Studies* 22, 59-76.

Hájek, P., [1988]: "Towards a probabilistic analysis of MYCIN-like expert systems", D. Edwards, N. Raun (eds.), "COMPSTAT 1988", Physica-Verlag, Heidelberg, 117-122.

Hájek, P., [1989]: "Towards a probabilistic analysis of MYCIN-like expert systems II," I. Plander (ed.), *Artificial Intelligence and Information Control Systems of Robots, North-Holland (in print).*

Hájek, P., Hájková, M., [1984]: "EQUANT - brief description and users manual", P. Hájek (ed.), *New enhancements in GUHA software*, Mathematical Institute of the Czech. Acad. Sci., Prague.

Hájek, P., Hájkov , M., Havránek, T., Daniel, M., [1989]: "The expert system shell EQUANT-PC - brief information", *Kybernetika 25, suppl.*, 4-9.

Hájek, P., Havránek, T., [1987]: "A note on independence assumptions underlying subjective Bayesian updating in expert systems", I. Plander (ed.), *Artificial Intelligence and Information Control Systems of Robots*, North-Holland, 41-47.

Hájek, P., Valdés, J.J., [1986]: "Algebraic foundations of uncertainty processing in rule-based systems", I. Plander (ed.), *Sistemy obrabotki znanij i izobrazhenij, vol. I*, Smolenice, 93-98.

Hájek, P., Valdés, J.J., [1987a]: "Algebraic foundations of uncertainty processing in rule-based systems", *AI'88, Ústav pro informační systémy v kultuře*, Praha, 9-14.

Hájek, P., Valdés, J.J., [1987b]: "Algebraic foundations of uncertainty processing in rule-based systems", "res. rep. 28", Mathematical Institute of Czech. Acad. Sci., Prague.

Havránek, T., [1989]: "Some notes to the algebraization of the Dempster-Shafer treatment of uncertainty", *Kybernetika 25, suppl.*, 10-14.

Havránek, T., Kouba, Z., [1988]: "On the expert system EPROP", *J. New Gener. Comput. Syst. 1*, 87-96.

Ivánek, J., Stejskal, B., [1988]: "Automatic acquisition of knowledge base from data without expert: ESOD", D. Edwards, N. Raun (eds.), *COMPSTAT 88*, Physica-Verlag, Heidelberg, 175-180.

Johnson, R.W., [1986]: "Independence and Bayesian updating methods", L.N., Kanal, J.F., Lemmer (eds.), *Uncertainty in Artificial Intelligence*, Elsevier, 197-202.

Lauritzen, S.L., Spiegelhalter, D.J., [1988]: "Local computations with probabilities on graphical structures and their application to expert systems", *J. Royal Statist. Soc. B 50*, 157-224.

Perez, A., [1984]: " 'Barycenter' of a set of probability distributions and its application in statistical decisions", T. Havránek, Z. Šidák, M. Novák (eds.), *COMPSTAT 84*, Physica-Verlag, Wien, 154-159.

Perez, A., [1987]: "The maximum entropy principle and the barycenter approach in knowledge integrations", *Third Joint USSR-Swedish International Workshop on Information Theory*, Sochi.

Perez, A., Jiroušek, R. [1985]: "Constructing an intensional expert system (INES)", *Medical Decision Making - Diagnostic Strategies and Expert Systems*, North-Holland, 307-315.

Shafer, G. [1976]: *A mathematical theory of evidence*, Princeton University Press.

Spiegelhalter, D.J., [1986a]: "A statistical view of uncertainty in expert systems", W. Gale (ed.), *Artificial Intelligence and Statistics*, Addison-Wesley.

Spiegelhalter, D.J., [1986b]: "Probabilistic reasoning in predictive expert systems", L.N. Kanal, J.F. Lemmer (eds.), *Uncertainty in Artificial Intelligence*, Elsevier, 47-67.

Valdés, J.J., [1987]: *Algebraic and logical foundations of uncertainty processing in Rule-based expert systems of artificial intelligence*, PhD thesis, Mathematical Institute of Czech. Acad. Sci., Prague.

Valdés, J.J., [1989]: "Partially ordered combining structures and multiple sources of knowledge evidences in rule-based systems", *Kybernetika 25, suppl.*, 15-19.

# Chapter 7

# Modelling Belief

J.B. Paris[1] and A. Vencovska[2]
*Department of Mathematics, Manchester University, Manchester. M13 9PL. England*

We criticise the common practice of treating *degree of belief* as subjective probability in explaining certain examples of human reasoning under uncertainty and suggest instead an explanation based on a mathematical model in which belief is a derived concept.

## 7.1 Introduction

The work described in this paper stems from attempting, with a colleague in Pathology [Paris & Rutherford - to appear], to produce an expert system for the differential diagnosis of cancer types on slide evidence alone.

Initially an expert pathologist was asked what communicable features on the slides he thought he was using in the diagnosis and how relevant he considered them to be. In this fashion the expert volunteered a number of rules. In response to direct querying these were

[1]Supported by a Science & Engineering Research Council Fellowship under the 'Logic for IT' Initiative

[2]Supported by a Science & Engineering Research Council Research Assistantship under the 'Logic for IT' initiative

augmented by further explanations about dependencies and base rates. Finally, in common with usual practice, these natural language statements were then converted into linear equations over the rationals involving the beliefs (Bel) of various propositions. This was carried out in consultation with the expert who was asked to quantify his belief on a scale [0,1], 1 corresponding to certainty, 0 to certainty in the negation and $\frac{1}{2}$ to no preference. Belief values strictly between zero and one in part arose because almost all the salient features (such as "blueness", "lined up cells") were vague and ill-defined and also because so-called rules were usually not definitive.

As an example of this conversion process the expert might suggest that "Blueness ($B$) is an indicator of Lymphoma ($L$) although not a very strong one". Asked to put a figure, on the scale mentioned above, to "is an indicator although not a very strong one" the expert might suggest, say, 0.6. His original statement might then be converted into the intermediate mathematical form "Belief in $L$ given $B$ is 0.6" and thence finally to

$$Bel(B \wedge L) = 0.6 \times Bel(B)$$ [3]

where $B$ is a propositional variable standing for Blueness and $Bel(B)$ is the expert's degree of belief on the above scale that an arbitrary, previously unseen slide, will on examination exhibit blueness. [For further details of our approach see [Paris & Vencovska 1989].]

The upshot of this conversion process then was that in a particular diagnostic situation we obtained a set $S$ of linear constraints on beliefs,

$$\sum_{j=1}^{n} \alpha_{ij} Bel(\theta_j) = \beta_i \qquad i = 1, ..., m$$

with $\alpha_{ij}$, $\beta_i$ rational, and the $\theta_j$ are sentences from some language in the sentential calculus. The diagnostic problem we then considered was how these constraints might determine the value(s) $Bel(\theta)$ of the belief in $\theta$ for some other sentence $\theta$. In short:

[3] There are other possible mathematical formulations of the expert statement, for example $Bel(B \rightarrow L) = 0.6$. However, the one given seems the most generally favoured interpretation and is the one we shall use in this chapter.

**Q:** "Given constraints $S$ what value(s) should be assigned to $Bel(\theta)$?".[4]

This question, and more generally the problem of formalizing and understanding possible modes of inexact or approximate reasoning, was the main stimulus for this chapter.

## 7.2 Approaches and Departures

Unfortunately it is very exceptional that direct linear algebra enables us to solve for $Bel(\theta)$ and we need to make further assumptions about the function $Bel$ in order to answer $Q$.

Various properties have been proposed, e.g.

- $Bel$ satisfies the axioms of probability; that is belief equals probability,
- $Bel$ is a Shafer Belief Function, (see [Shafer 1976])
- $Bel$ is a valuation in the Lukasiewicz Interval (Fuzzy Logic),
- $Bel$ is a confidence factor in the sense of, e.g. Prospector, and hence takes intermediate values in an ordered abelian group on (0,1), (see [Hájek 1985]) etc.

Each approach has its supporters, even fanatics. However, by far the most popular assumption is that belief equals probability and we shall now briefly criticise this assumption within this context. This focusing of attention should not be taken to imply that we are particularly enamoured of any of these other alternatives.

An immediate difficulty in assuming $Bel$ to satisfy the axioms of probability is that, in our experience, it not infrequently renders the constraints given by the expert trivially inconsistent. For example the expert would volunteer constraints like

$$Bel(A) = 0.8, \quad Bel(A \wedge B) = 0.8 \times Bel(A), \quad Bel(B) = 0.4$$

[4]The more practically useful generalization of this question, that of giving a value to $Bel_s(\theta)$, the belief in $\theta$ given an observation $s$, will be considered later.

which imply $Bel(A \wedge B) = 0.64 > Bel(B)$ contrary to the axioms of probability. One could of course argue that we can easily detect such minor inconsistencies and they can be corrected with the acquiescence or re-education of the expert. However, the problem will not disappear so easily as the following result shows.

**Theorem 1** *([Paris & Vencovska 1989]). With Bel satisfying the axioms of probability the problem of deciding if the set of constraints*

$$Bel(\theta_i) = \beta_i \quad i = 1, ..., m$$

*is consistent is NP-complete and hence (apparently) computationally infeasible. Indeed we may take the $\theta_i$ here to be conjunctions of $\leq 3$ propositional variables or their negations.*

We hasten to point out that in expressions like these equality is taken to be approximate in the sense that by *consistent* we only mean that there are solutions which are arbitrarily close to satisfying the given constraints. [In fact in this case such a distinction does not matter although it will in later examples.]

Of course, these so called inconsistencies do not appear to bother the expert in his reasoning.

A second anomaly with the figures we received from experts was that they sometimes appeared to "miss" perfectly good rules. For example the expert might state that $A$ was a strong indicator of $B$ and give figures

$$Bel(A) = 0.6, \quad Bel(A \wedge B) = 0.8 \times Bel(A), \quad Bel(B) = 0.5$$

but be completely oblivious of the fact that with these figures and belief as probability, $\neg A$ is an even stronger indicator of $\neg B$. It seems perhaps strange that if the expert really is using rules he can be so unobservant.

Of course, these examples were taken from our very limited experience. However, there are serious studies which support the thesis that human reasoning is not carried out according to the laws of probability, at least in the normal way of interpreting them (see for example [Kahneman & Tversky 1982]).

Returning now to question $Q$ and assuming $S$ to be consistent the assumption of Belief as probability is still usually insufficient to uniquely determine $Bel(\theta)$ and further assumptions are required if such is to be achieved. In fact this is possible as the following paraphrased version of the main result of [Paris & Vencovska, to appear] shows.

**Theorem 2** *Under the assumption that Bel satisfies the axioms of probability there are "common sense" principles applying to the* process

$$\textit{Consistent } S \longmapsto \textit{value of } Bel(\theta)$$

*which force $Bel(\theta)$ to be uniquely determined. This unique value is the probability of $\theta$ for that probability distribution satisfying $S$ which has the maximum entropy.*

An example of such a common sense principle[5] is

**P1:** For fixed consistent $S$ but varying $\theta$ the values given to the $Bel(\theta)$ should all be consistent with each other and with $S$.

However, despite the crystallization this theorem seeds we feel it does little to strengthen the case for belief as probability. We mention two criticisms of the use of this so called Maximum Entropy Inference Process (ME) in this situation.

The first, and this is true of almost any process in fact, is that $Bel(\theta)$ can depend non-trivially on how deeply we analyse properties not directly mentioned in $\theta$, more precisely after what degree of analysis we decide to represent a property simply as a propositional variable. For example, replacing "green" by "blue & yellow" in $S$, where "blue" and "yellow" did not previously appear in $S$ or $\theta$, can alter the value the process would give to $\theta$, even if $\theta$ does not mention "green".

A second criticism, as was shown in [Paris & Vencovska 1989], is that the problem of computing ME values even approximately is, assuming $P \neq NP$, infeasible. In fact if the generally held view in computational complexity that $RP \neq NP$ is assumed (see [Valiant & Vazirani 1986]) then infeasibility is a much more general phenomenon as the following theorem shows.

---

[5]One could criticise the assumption of "common sense" principles in that it is based on our beliefs about beliefs rather than by direct experience. This point will emerge again later.

**Theorem 3** *[Maung & Paris]. Assume that $RP \neq NP$ and that Bel satisfies the axioms of probability. Let $0 \leq a < \frac{1}{2} < b$. Then there is no (random) computational process which runs in polynomial time and which on input of a consistent set $S$ of constraints and a proposition variable $p$ outputs a number $\nu$ such that, (with probability $\geq b$),*

$$S + [\nu - a \leq Bel(p) \leq \nu + a]$$

*is consistent.*

Of course this fails if $a = \frac{1}{2}$ - take $\nu = \frac{1}{2}$! The essence of this result then is that in general, given constraints $S$, we cannot (tractably) hope to find any non-trivial bounds on $Bel(p)$, even when $p$ is just a propositional variable. Furthermore $S$ may be assumed to consist simply of constraints of the form $Bel(\theta_i) = \beta_i$ where $\theta_i$ is a conjunction of at most three propositional variables or their negations.

It could of course be argued that this is only an asymptotic result applying to each particular computational process but manifesting itself perhaps only for very large $S$ and hence is not applicable to a real world expert system. That may indeed be so, but experience to date does not suggest that the computational difficulties inherent in, say, natural NP-complete problems are only manifest "beyond the horizon".

In order to bring out a point later we suggest a possible cause for the difficulty in finding a consistent value for $Bel(p)$. The set $S$ of constraints involves sentences made up from some propositional variables $p_1, ..., p_k$ say. In general the number of constraints in $S$ will be bounded by some small polynomial in $k$. However, to find a probability function $Bel$ to satisfy $S$ seems to require $S$ to be re-expressed and solved as a set of linear equations in the $2^k$ variables $Bel(\wedge_{i=1}^k \pm p_i)$.

Thus the solving for $Bel$ is taking place not at the level of the $k$ propositional variables but at the level of the $2^k$ atoms of the corresponding Lindenbaum algebra. Notice that such a method of solution does not seem to correspond to the human expert's reasoning - that appears to take place at the higher level of the propositional variables.

## 7.3 Modelling Belief

Although we have concentrated on the assumption of Belief as probability, versions of theorems 1 and 3 also hold for Shafer belief functions and for fuzzy valuations on the Lukesiewicz Interval, so these approaches fare no better as regards computational tractability.

Nevertheless, in human reasoning under uncertainty the problems of inconsistencies and infeasibility do not appear to be significant, which suggests there might be still other approaches to inexact reasoning.

Each of the approaches mentioned above make assumptions about what properties belief should have. However, to our minds these assumptions are far from self evident, indeed we would be wary of any assumptions about belief derived solely from intuition and introspection.

But how else is there to understand belief except by treating it as primitive and then attempting to axiomatize it? One seemingly neglected alternative approach is to treat belief as a derived concept, defined in terms of simple concrete mathematical models (but see [Halpern 1989]). After all it appears that in the real world our beliefs are largely derived from experience, so any properties that beliefs possess should be demonstrable within the context of experiences.

We do not claim that this will give a definitive answer to the question of what belief is (even assuming there is a definitive answer). However, it will surely give new insights and suggest new possibilities, unimpeded by conditioned reactions.

The idea then is simply to give a model or meaning to a set of constraints $S$ of the form which might have been given by the expert. Once we have this model the relationship between $Bel(\theta)$ and the constraints will simply drop into our lap.

## 7.4 A First Model

In [Paris & Vencovska 1989] we considered one such model (phrased in terms of slide diagnosis as above) and we now briefly recall it. In this model we assumed that the expert pathologist had in his memory a large set $N$ of slides, each of which was completely known in terms of which properties it did or did not possess, and we defined

$(Bel(\theta) = \alpha) \equiv$ (proportion of slides in $N$ satisfying $\theta = \alpha$),
etc.

Here as usual equality is approximate. In this setting question $Q$ becomes

"Given that $N$ "satisfies" the constraints $S$ determine the proportion of slides in $N$ satisfying $\theta$."

Unfortunately of course we do not know $N$, only that $S$ is satisfied. However, (see [Paris & Vencovska 1989] for the precise result),

**Theorem:** For $n$ large and for almost all sets $N$ with size $n$ satisfying S,

(proportion of slides in $N$ satisfying $\theta$)

$= \text{ME value of } \theta \text{ given the constraints } S.$

That is, almost all expert pathologists who gave $S$ would be in close agreement about their belief in $\theta$.

So, with this model, belief does come out as probability, and indeed this result is a further argument for identifying reasoning under uncertainty in this context with the Maximum Entropy Inference Process. Moreover the problems with computational complexity, for the expert at least, seems miraculously to have disappeared! The expert simply takes a relatively small random sample from $N$ and checks the proportion of these satisfying $\theta$.

Well, as we pointed out in [Paris & Vencovska 1989], this model has a major practical deficiency in that simply to avoid the expert giving belief 0 to consistent statements, let alone clustering around ME, requires in general the size of $N$ to be approximately exponential in the size of $S$, and hence practically unattainable in all except trivial cases.

In short this model requires the expert to have had an immense amount of experience coupled with a quite phenomenal memory in which to store it all! Clearly from memory capacity considerations alone it seems more reasonable that the expert's view of what such terms as "blueness" mean is based, not on a subset of some set of stored, perfectly known slides, but on remembered fragments of rather few slides.

A second, and closely related, criticism of this first model is that it assumes that properties like "blueness" either hold or do not hold of a particular slide and that in theory (and in this idealized model of belief) any slide can be perfectly known. Thus the fact that a pathologist might feel uncertain about whether or not he was seeing "blueness" or "linear cell arrangement" would be explained in terms of him being able to see only a blurred picture of something which definitely was, or was not, there.

But another view of this latter situation is that the expert sees all that there is to be seen. It is the possibility the expert has of re-interpreting his view of the slide which causes these uncertainties. That is, the expert himself generates the additional knowledge, or as we shall call them, scenarios, rather than the additional knowledge objectively existing.

Furthermore we could imagine this additional knowledge being generated by forming combinations of the above mentioned remembered fragments consistent with what is seen on the slide. Thus this process of knowledge generation rather resembles modelling from Lego, combining versatility with low storage space requirement. Notice that our first model above could be viewed as a very special case of such modelling where all the models are already built and on the shelf. With these ideas in mind we now tentatively suggest a second mathematical model of the expert's given constraints $S$. We remark that these ideas are still very much in the melting pot and that this model should be viewed as a suggested direction rather than definitive.

## 7.5 A Second Model

Before describing this model we wish to stress that it is only a *model*. We are most definitely not claiming that it is the way human beings think, only that it is a way some entities (who might perhaps consider themselves intelligent) *might* conceivably act. To ensure there is no confusion on this point we shall refer to these entities as "Ents".

The idea is that when questioned about his belief in some primitive proposition (i.e. one represented by a propositional variable in the formulation of $S$) the Ent simply guesses or imagines scenarios consistent with what he currently takes to hold true and in which the proposition is decided. His numerical statement of belief in the proposition is simply the proportion of the imagined scenarios in which it is decided positively.[6] [Notice that the first model falls under this scheme - there the Ent guesses scenarios with complete information.]

Compound sentences are read from left to right, extending the currently held scenario whenever there is any doubt. Further, this same procedure is adopted whether the Ent is answering a general question about his beliefs or making, say, a specific diagnosis, although in general in this latter case an observation $s$ would initially be taken as given. [We shall, without loss of generality, assume that in answering general questions about beliefs nothing is taken as initially given.]

Not a particularly complicated model then, although as we shall see it has some unexpected properties.

We now give a more precise definition which includes one additional simplification. Namely, an imaginable scenario is assumed to have the same potential for being imagined at any point at which it is consistent with the currently held truths (i.e. scenario) and decides the propositional variable currently being decided. [A full account of this work will appear in [Paris & Vencovska c]].

## 7.6 Definitions

A scenario $s$ is a consistent subset of the set of $\pm p$ with $p$ a propositional variable in some finite language for the sentential calculus (SC).

An Ent is a set $E$ of scenarios with strictly positive potentials $z_e$, $e \in E$ such that for any scenario $s$ and any propositional variable $p$ not decided by $s$ (i.e. $p \notin s$ & $\neg p \notin s$ - also expressed as $\pm p \notin s$) there is a scenario $t$ in $E$ consistent with $s$ and deciding $p$.

---

[6] In practice the Ent would generate only a few scenarios in the time he allowed himself to answer the question. This then accounts for Ents sometimes revising their opinions after further considerations. For mathematical simplicity we shall idealize by taking the limit of this process.

Here $z_e$ is the potential for $e \in E$ to be guessed or imagined.

Given a scenario $s$, Ent $E$ and a propositional variable $p$ the belief, $Bel_s(p)$, that $E$ has in $p$ given $s$, is defined as follows:

$$\text{if } p \in s,\ Bel_s(p) = 1;\ \text{if } \neg p \in s,\ Bel_s(p) = 0.$$

Otherwise

$$Bel_s(p) = \text{Probability of guessing from}$$

$$\{e \in E \mid e \cup s \text{ is consistent \& } e \text{ decides } p\}$$

an $e \in E$ with $p \in e$,

$$= \frac{\sum\{z_e \mid e \in E, e \cup s \textit{ consistent} \ \&\ p \in e\}}{\sum\{z_e \mid e \in E, e \cup s \textit{ consistent} \ \&\ \pm p \in e\}}.$$

Belief in $\neg p$ is defined in the obvious way by replacing $p$ everywhere by $\neg p$. Conjunctions etc. are handled by repeatedly guessing scenarios whenever necessary, reading from left to right. To make this precise we need to define an additional function $G$ on pairs of scenarios. [The reader may wish to skip straight to the example which follows since that makes it clear how conjunction is treated.]

For $s$ a scenario which decides $p$ set

$$G_p(s, s') = \begin{cases} 1 & \text{if } p \in s,\ s = s' \\ -1 & \text{if } \neg p \in s,\ s = s' \\ 0 & \text{if } s \neq s' \end{cases}$$

If $s$ does not decide $p$ set

$$G_p(s, s') = \frac{\sum\{z_e \mid e \in E,\ s' = s \cup e\}}{\sum\{z_e \mid e \in E,\ s \cup e \textit{ consistent} \ \&\ \pm p \in e\}}$$

if $p \in s'$,

minus this if $\neg p \in s'$,

and otherwise $G_p(s, s') = 0$.

Now for sentences $\theta, \varphi$ of the language define inductively

$$G_{\neg\theta}(s, s') = -G_\theta(s, s')$$

$$G_{(\theta\wedge\varphi)}(s, s') = \begin{cases} \sum_t G_\theta(s,t)G_\varphi(t,s') & \text{if } s' \vdash^{sc} \theta \\ G_\theta(s,s') & \text{if } s' \vdash^{sc} \neg\theta \\ 0 & \text{otherwise} \end{cases}$$

Disjunctions, implications etc. can be defined similarly or, equivalently, via the definitions

$$(\theta \vee \varphi) = \neg(\neg\theta \wedge \neg\varphi), \ (\theta \rightarrow \varphi) = (\neg\theta \vee \varphi)$$

Belief, for an Ent $E$, given an observation or scenario $s$, is now defined by

$$Bel_s(\theta) = \sum\{| \, G_\theta(s,s') \, | \ \big| s' \vdash^{sc} \theta, s' \text{ a scenario}\}$$

We set $Bel = Bel_\emptyset$.

These definitions might appear rather contorted. However, the following example will, we hope, make everything clear.

**Example:** Let $E$ be the Ent

| Scenarios: | $\{p\}$, | $\{q\}$, | $\{p, \neg q\}$, | $\{\neg p\}$ |
|---|---|---|---|---|
| Potentials: | $\frac{1}{4}$ | $\frac{1}{4}$ | $\frac{1}{4}$ | $\frac{1}{4}$ |

Using the above definition, with $s = \emptyset$, gives

$$\begin{aligned} Bel(p) &= \frac{\text{Sum of the potentials of } \{p\}, \{p, \neg q\}}{\text{Sum of the potentials of } \{p\}, \{p, \neg q\}, \{\neg p\}} \\ &= \frac{\frac{1}{4}+\frac{1}{4}}{\frac{1}{4}+\frac{1}{4}+\frac{1}{4}} = \frac{2}{3} \end{aligned}$$

$Bel(p \wedge q)$ is computed by first guessing a scenario (according to its potential) which decides $p$, so for example this lands us at $\{p\}$ with probability $\frac{\frac{1}{4}+\frac{1}{4}}{\frac{1}{4}+\frac{1}{4}+\frac{1}{4}} = \frac{2}{3}$, and then, if, as is the case here, $p$ holds at this scenario, picking, if necessary, and according to potentials a scenario deciding $q$ and consistent, in this case, with $\{p\}$. So for example at $\{p\}$ and deciding $q$ the Ent $E$ would have probability $\frac{1}{2}$ of picking $\{p, \neg q\}$ and at this point $p \wedge q$ is decided of course. $Bel(p \wedge q)$ is the probability of ending at a scenario which decides $p \wedge q$ positively. Thus $Bel(p \wedge q)$ is computed via the diagram

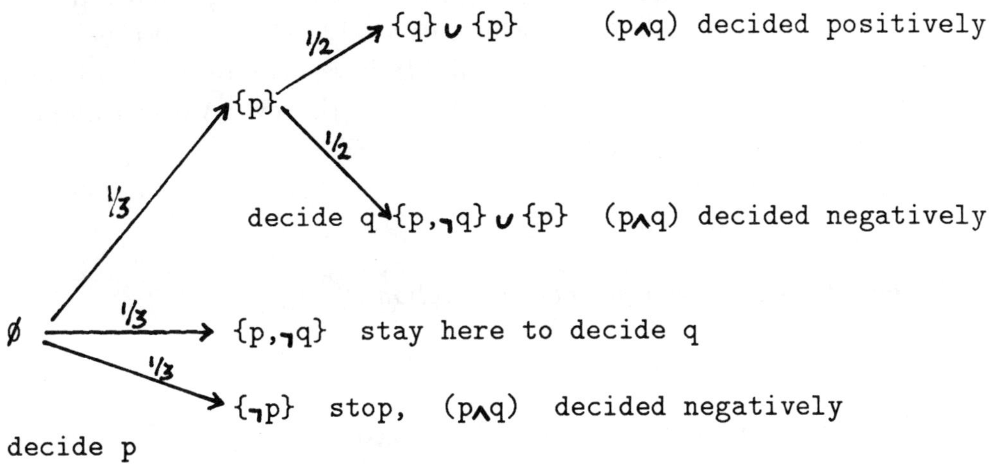


$Bel(p \wedge q) =$ Prob. of ending in a scenario with $(p \wedge q)$ true $= \frac{1}{3} \times \frac{1}{2} = \frac{1}{6}$. Notice that the Ent does not need to draw this diagram, he can make a good estimate of this figure by simply running the process a few times. Furthermore this applies in computing any $Bel_s(\theta)$. For the variance in randomly deciding $\theta$ in this way, say, $n$ times and averaging is $n^{-1}Bel_s(\theta)(1 - Bel_s(\theta)) \leq 1/4n$ and, independent of the complexity of $\theta$, taking $n = 1/4\varepsilon^3$ ensures with probability at least $1 - \varepsilon$ an estimate of $Bel_s(\theta)$ within $\varepsilon$. Thus in computational complexity terms estimating $Bel_s(\theta)$ is completely feasible. Indeed we would suggest that a sensible Ent would classify his scenarios according to the propositional variables they decided so that when asked to pick a scenario deciding $p$, say, he could do it in one step without having to randomly keep picking arbitrary scenarios until one was found deciding $p$. [This seems reasonable, after all, if we asked you to imagine, say, a lion, you wouldn't run a whole series of mental pictures until you came across one containing a lion!] This could further reduce the running time in the (likely) event that $\theta$ only mentioned a fraction of the propositional variables mentioned in $E$.

A possibly surprising, feature here is that for this Ent,

$$Bel(q \wedge p) = \frac{1}{4} \neq \frac{1}{6} = Bel(p \wedge q).$$

This could be considered surprising in that in making a natural generalisation of our first model which gave ME we have lost belief as probability. We would regain probability if we required each scenario in $E$ to decide all the propositional variables. However, we have already put forward objections to this.

In fact in this modelling commutativity of conjunction and belief as probability are equivalent:

**Theorem 4** *$Bel_s$ is a probability function iff*

$$\forall \theta, \varphi \; Bel_s(\theta \wedge \varphi) = Bel_s(\varphi \wedge \theta).$$

So it seems that for Ents the order of a conjunction may be relevant, perhaps an Ent would not even see the sentences

"She got married and had a baby"

"She had a baby and got married"

as saying exactly the same thing. Similar remarks apply with disjunction in place of conjunction.

From the outside then an Ent's degree of belief function *Bel* may seem to have very different properties from those normally assumed to hold for us (and hence expressed in axiomatizations of belief). However, Ents believe, like us, that conjunction *is* commutative. Precisely:

**Theorem 5** $s \vdash^{sc} \Psi \iff \forall Ents,\ Bel_s(\Psi) = 1$

In particular with $\Psi = ((\theta \wedge \varphi) \longleftrightarrow (\varphi \wedge \theta))$ all Ents are certain that $(\theta \wedge \varphi) \longleftrightarrow (\varphi \wedge \theta)$ is true. Indeed they might say it was "common sense"!

Furthermore Ents are similarly convinced that proofs in the sentential calculus preserve truth, precisely:

**Theorem 6** *For an Ent E if* $Bel_s(\theta_i) = 1$ *for* $i = 1, ..., n$ *and* $\theta_1, \theta_2, ..., \theta_n \vdash^{sc} \theta$ *then* $Bel_s(\theta) = 1$.

Our next two theorems answer, albeit rather messily, the question of what pairs of sentences must always have (for each Ent) the same $G$ function and which pairs always have the same belief.

**Theorem 7** *For all Ents,* $G_\eta = G_\tau$ *iff* $\eta \sim \tau$ *is derivable for any relation* $\sim$ *on sentences satisfying*

(i) If $\theta_1 \sim \Psi_1$, $\theta_2 \sim \Psi_2$ then $(\theta_1 \wedge \theta_2) \sim (\Psi_1 \wedge \Psi_2)$, $(\theta_1 \vee \theta_2) \sim (\Psi_1 \vee \Psi_2)$, and $\neg\theta_1 \sim \neg\Psi_1$.

(ii) $\sim$ is an equivalence relation.

(iii) $\theta \sim \neg\neg\theta$ $\qquad \theta \wedge (\varphi \wedge \Psi) \sim (\theta \wedge \varphi) \wedge \Psi$

$\neg(\theta \wedge \Psi) \sim \neg\theta \vee \neg\Psi \qquad \theta \vee (\varphi \vee \Psi) \sim (\theta \vee \varphi) \vee \Psi$

$$\theta \wedge (\varphi \vee \lambda) \sim (\theta \wedge \varphi) \vee (\theta \wedge \lambda)$$

$$(\theta \vee \varphi) \wedge \lambda \sim (\theta \wedge \lambda) \vee (\neg\theta \wedge (\varphi \wedge \lambda))$$

$$\theta \wedge \neg\theta \sim \neg\theta \wedge \theta$$

$$(\theta \wedge \neg\theta) \wedge \varphi \sim (\theta \wedge \neg\theta) \qquad \theta \wedge (\neg\theta \vee \varphi) \sim \theta \wedge \varphi$$

$$\theta \wedge (\Psi \wedge \theta) \sim \theta \wedge \Psi \qquad \theta \vee (\theta \wedge \Psi) \sim \theta \quad (\Psi \text{ may be absent})$$

$$(\theta \wedge \Psi) \wedge (\varphi \wedge \neg\varphi) \sim (\theta \wedge \Psi) \wedge (\varphi \wedge \neg\theta) \quad (\Psi \text{ may be absent})$$

**Theorem 8** *For all Ents,* $Bel(\eta) = Bel(\tau)$ *iff* $\eta \approx \tau$ *is derivable for any relation* $\approx$ *on sentences satisfying*

(iv) if $\theta \sim \varphi$ is derivable on the basis of (i), (ii), (iii) of theorem 7 then $\theta \approx \varphi$.

(v) For $\varphi$ a tautology $\theta \approx \theta \wedge \varphi \approx \theta \vee \neg\varphi$

(vi) $\approx$ is an equivalence relation.

Let us now return to the criticisms we levelled earlier against belief as probability in the situation of the expert pathologists statements.

For the Ent pathologist:

(i) The cited typical simple inconsistency is no longer an inconsistency.

(ii) "Missing" rules no longer need hold.

(iii) In "real situations" we can now test for consistency in polynomial time.

Let us explain what we mean in (iii). By the result,

**Theorem 9** *$Bel(\theta) = 1$ is satisfiable iff $\theta$ is satisfiable in SC,*

we still cannot expect to test for consistency, in polynomial time, arbitrary finite sets $S$ of linear constraints in, say, $Bel(\theta_i)$ $i = 1, ..., m$. However, the reason for the (apparent) intractability here is due to the individual $\theta_i$ becoming very complicated - which is not a feature of real, practical examples. In fact:

**Theorem 10** *Let $k$ be fixed and let $S$ be a set of constraints in $Bel(\theta_i)$ $i = 1, ..., m$ with the $\theta_i$ containing at most $k$ connectives. Then it is possible to test the consistency of $S$ in polynomial time (consistency in a language possibly containing propositional variables not occurring in $S$).*

In fairness we should add that theorem 10 is something of a bonus since feasible testing of consistency appears irrelevant from the Ents' point of view.

Returning now to our earlier list of criticisms we now notice that

(iv) This model of the Ent expert does not make unrealistic demands on his time and space.

In fact there is a bonus here that we again would have had no reason to expect:

**Theorem 11** *Let $k$ be fixed and let $S$ be a consistent set of constraints in $Bel(\theta_i), i = 1, ..., m$ with the $\theta_i$ containing at most $k$ connectives. Then we can describe Ents which arbitrarily closely satisfy these constraints and further we can give approximate values to $Bel_s(\theta)$, in random polynomial time so as to satisfy the principle $P1$. (These Ents may be defined for a larger language than just that of $S$ and $\theta$.)*

The reason for this, unlike the case of probability, is that now we can find solutions at the level of the propositional variables rather than at the level of the $\wedge_{i=1}^{k} \pm p_i$.

## 7.7 Relevance of this Model to Expert Systems

Returning to the original motivation for this chapter, the interpretation of the expert pathologist's belief statements and subsequent development of an expert system, we now consider the case where the pathologist is also an Ent. Let us suppose that in response to our questioning he has volunteered belief statements.

$$Bel(\theta_i) = \alpha_i \;\; i = 1, ..., m$$

- as indicated in [Paris & Vencovska 1989], in practice, we can usually assume that his general belief statements take this simplified form and indeed further, that the $\theta_i$ are simple. Then assuming these constraints are consistent (and by theorem 10 we can check consistency) we can, by theorem 11 (to an arbitrarily good approximation), describe an Ent to satisfy these constraints and thence give (approximate) values to $Bel_s(\theta)$ so as to satisfy the principle $P1$.

In this fashion then we can construct an expert system for answering questions of the form $Bel_s(\theta) = ?$ for given $s$ and $\theta$ which is consistent with our interpretation of the Ent expert's given beliefs. This is rather better than just answering our original question $Q$ in that it enables us to directly take into account observational datum $s$ in the resulting expert system. However, there is absolutely no justification for believing it to be any sort of a global approximation to the actual expert Ent on whose constraints it was based. For whilst this interpretation of the belief statements made by the Ent expert may to some extent avoid the inconsistencies arising under the assumption that belief equals probability (we do not claim that all inconsistencies are removed) it has the disadvantage for the builders of expert systems that these given belief statements may now provide almost no information. For example consider the classic situation of wishing to estimate the belief in $q$ having been told by an Ent expert that his belief in $p$ is $\alpha$ and his belief in $q$ given $p$ is $\beta$ $(0 < \alpha)$. There are various possible formulations of this second statement, for example $Bel(p \to q) = \beta$ or $Bel_p(q) = \beta$ or $Bel(p \wedge q) = \beta\ Bel(p)$, (of which we favour the last when $p$ is hypothetical, as here). However, no matter which of these is taken, without further assumptions the expert's statements give no non-trivial information about $Bel(q)$ as can easily be seen. For example the Ent with potentials

$$\begin{array}{cccc} \{p,q,r\}, & \{p,r\}, & \{\neg p\}, & \{\neg q, \neg r\} \\ 0.001 & 0.089 & 0.01 & 0.799 \end{array}$$

satisfies

$$\text{Bel(p)} = 0.9$$

$$\text{Bel(p}\wedge\text{q)} = 0.9 = \text{Bel(p)}$$

$$\text{Bel(q)} = .00125$$

$$\text{Bel(p}\to\text{q)} = 1,$$

whilst for the Ent

$$\begin{array}{ccc} \{p\}, & \{\neg p, \neg q\}, & \{q\} \\ 0.9 & 0.09 & 0.01 \end{array}$$

$\text{Bel(p)} = 0.9$, $\text{Bel}_p\text{(q)} = 1$, $\text{Bel(q)} = 0.1$

So what is required here it seems is some sort of justification for picking a particular solution from the multitude of possible solutions. Various heuristics can be applied at this point. For example we might pick an approximating Ent which is as "simple" as possible according to some measure of simplicity - for example that the Ent should consist as much as possible of small scenarios, on the grounds that this generally increases flexibility whilst minimising storage space. Alternatively we could require the Ent to be as quick as possible within the mode of operation formulated earlier.

A criticism of such justifications is that they tend not to determine unique solutions even in quite simple cases.

A much more fundamental criticism however is that whilst it might seem to be desirable for the Ent to be so efficient we know of no justification for assuming that (if he really existed!) he would be. Of course it may be that there is a simple, practical, procedure that an Ent could adopt to rearrange himself to some canonical simple form whilst maintaining approximately the same beliefs. However, we know of no such procedure.

An alternative approach to heuristic assumptions is to consider how the Ent expert might have evolved or developed from a state of ignorance and to use this to limit the class of solutions of the constraints which are prospective approximations. For example we could suppose that Ents acquire their knowledge (i.e. scenarios) as follows. Initially the novice Ent randomly assigns microscopic non-zero potentials to all singleton scenarios $\{\pm p_i\}$. Thereafter he accepts, from his experiences, a stream of scenarios which he stores increasing $z_e$ by 1, say, every time he receives scenario $e$. In this way he eventually becomes an expert.

This description of course completely sidesteps the central question of how he decides what to accept. What can, however, be quite reasonably asserted (see [Paris & Vencovska c]) is that accepting large scenarios increases accuracy of judgements but at the same time enormously increases the time required to graduate from novice to expert. It seems then, in general, that accepting small scenarios is more practical, whilst some accuracy can be maintained by concentrating on scenarios which decide relevant features. Presumably then the novice would benefit by having a tutor point out what was relevant during the training process.

Notice that if an expert Ent did evolve in this way then there would it seems be no reason why it should satisfy some heuristic requirement per se. Furthermore, the knowledge the expert would end up with would be uniquely his own, although of course one might expect some agreement between Ent experts since they would have been educated by sampling from the same population. Notice, however, that there is no reason at all to assume that they are using the same $p_i$, i.e. the same features.

Indeed this possibility would provide a ready explanation of why some Ent experts might be better than others [and indeed agrees with the opinion of several human experts we spoke with as to why they were successful]. At the same time this poses further difficulties for developing successful Ent approximation to an Ent expert. For according to our formulation, say in a specific slide diagnosis situation, the Ent expert's conclusion is based on, for example, the value of $Bel_s$ (*Lymphoma*) where $s$ is what he *actually* sees on the slide. However, there is no reason why these features, as the expert sees them, should be communicable. For example one of them could be 'Lymphoma'!

To sum up, whilst we believe that Ents provide a practical model of intelligent beings capable of generating beliefs, in a world of Ents the development of expert systems aimed at imitating 'Ent intelligence' would face some non-trivial difficulties.

## References


Hájek, P., [1985]: "Combining functions for certainty degrees in consulting systems", *I.J. Man-Machine Studies 22.*

Halpern, J.Y., [1989]: "The relationship between knowledge, belief and certainty: preliminary report", *Proceedings of the Windsor Uncertainty Management Conference.*

Kahneman, D., & Tversky, A., [1982]: "Judgement under uncertainty: Heuristics and Biases". In Kahneman, Slovic & Tversky, (eds.) *Judgement under Uncertainty: Heuristics and Biases*, Cambridge University Press.

Maung, I., & Paris, J.B., [in preparation]: *A note on the infeasibility of some inference processes.*

J.B. Paris & A. Vencovska, [1989]: "On the applicability of maximum entropy to inexact reasoning", *International Journal of Approximate Reasoning*, Vol 3, No 1.

Paris, J.B., & Vencovska, A., b [to appear]: "A note on the inevitability of maximum entropy", in *the International Journal of Approximate Reasoning.*

J.B. Paris & J.J. Rutherford, [to appear]: *PARUT - An expert sys-*

*tem for the differential diagnosis of difficult Lymphoma/Epithelial Tumour slides*, in preparation.

Paris, J.B., & Vencovska, A., c [in preparation]: *A model of belief.*

Shafer, G., [1976]: *A mathematical theory of evidence*, Princeton University Press.

Valiant, L.G., & Vazirani, V.V., [1986]: "NP is as easy as detecting unique solutions", *Theoretical Computer Science,* Vol. 47.

# Chapter 8

# BACON Bytes Back

Peter Gibbins[1]
*Cranfield Information Technology Institute, Fairways, Pitfield, Kiln Farm, Milton Keynes MK11 3LG, England*

The philosophical reputation of Francis Bacon (1561-1626) – essayist, statesman, courtier, philosopher, bribe-taker, variously known as Sir Francis Bacon, Viscount St Alban, and plain *Verulam* – has experienced a recovery in recent years.

The Bacon boom has two sources – one philosophical and one computational. But they appear to pull in opposite directions. On the one hand, the philosophical rehabilitation of Bacon makes him out to be a more sophisticated philosopher of science than has been thought – less inductivist and more hypothetico-deductivist than the traditional view of Bacon admits. On the other hand, computational Baconianism, in the form of a suite of computer programs[2] which discover scientific theories when given data, seems to embody a traditional and pure inductivism.

In this chapter I shall argue that – despite appearances to the contrary – computational Baconianism is in one respect an implementation

[1]With thanks to Paul Gilbert, Peter Urbach and Dyson Wilkes for their comments on earlier drafts of this chapter.

[2]Described in Langley P, Simon HA, Bradshaw GL and Zytkow JM [1987] Scientific Discovery - Computational Explorations of the Creative Process. Cambridge, Mass: The MIT Press.

of Baconianism of the sophisticated rather than the traditional sort. I shall also argue that computational Baconianism suffers from one of the errors – the *anticipation of nature* – that Bacon himself warned against.

There is also a much larger philosophical question. Is computational Baconianism an adequate philosophy of science? On the basis of the prototype implementation of one computational Baconian computer program described in this chapter, I claim that the model of discovery built into computational Baconianism is *too algorithmic*, and insufficiently "intelligent", to supply even the beginnings of a theory on which an adequate conception of science may be based.

# Section 1

## 8.1 The Standard View of Francis Bacon

Peter Urbach, in a paper[3] and a book[4], has argued that Francis Bacon was by no means the hopelessly naive inductivist that numerous commentators have made him out to be, and that the *standard view*[5] of Bacon – the view that Bacon thought that his methodology afforded *a purely mechanical means of deriving demonstrably true scientific theories from the most superficial empirical data*, and that such theories are preferable to mere speculation – is a gross distortion.

The picture of Francis Bacon that Karl Popper paints is standard, in Urbach's sense. According to Popper

> "[t]he framework of Bacon's theory of induction is this. He distinguishes in the *Novum Organum* between a true method and a false method. His name for the true method, 'interpretatio naturae', is ordinarily translated by the phrase

[3]Urbach [1982].

[4]Urbach [1987].

[5]Otherwise called the Ellis-Hesse interpretation, or again, the infallible - mechanical interpretation. See Urbach [1987] p. 14 where these alternative descriptions of the standard view of Bacon are justified.

> 'interpretation of nature', and his name for the false method, 'anticipatio mentis', by 'anticipation of the mind'."[6]

The phrase "interpretation of nature" suggests

> "the avoidance of all interpretation in the modern sense, and it should not contain, more especially, any suggestion of an attempt to interpret what is manifest in nature in the light of non-manifest causes or hypotheses; for this would be an anticipatio mentis in Bacon's sense. (It is a mistake, I think, to ascribe to Bacon the teaching that hypotheses or conjectures may result from his method of induction; for Baconian induction results in certain knowledge rather than conjecture.)"[7]

Of his – Popper's – principle that satisfactory scientific theories must transcend the empirical instances which give rise to them, Popper writes

> "[h]ere we have a methodological principle which stands in direct contradiction to all postivistic and naively empiricist [or inductivist] tendencies. It is a principle which demands that we should dare to put forward hypotheses that open up, if possible, new domains of observations, rather than those careful generalizations from "given" observations which have remained [ever since Bacon] the idols of all naive empiricists."[8]

Popper's Bacon is thus a naive inductivist who dogmatically opposed deep, "bold" theories, and who thought that science could arrive at demonstrably true theories merely by employing induction by enumeration. Bacon therefore provides the foil against whom Popper can exhibit his own philosophy of science as properly cognizant of the uncertain and the creative in scientific activity.

[6] Popper [1968] p. 13.

[7] Popper [1968] p. 14.

[8] Popper [1972] p. 355. The phrase in parentheses is in Popper's text.

## 8.2 Francis Bacon as a Proto-Popperian

The Bacon that emerges from Urbach's treatment is quite different from Popper's. Urbach's Bacon is, surprisingly, a proto-Popperian, if not an entirely consistent one. Thus, according to Urbach, although Bacon did oppose what he called *the anticipation of nature*, he did not oppose scientific speculation as such, but only the *rigid adhering* to hypotheses which would then act as blocks to, or substitutes for, further empirical research[9]. In fact, Bacon's main criticism of the method of anticipation was that it was *not* sufficiently speculative, in the sense that it was founded on, and adhered to, observation of a small number of particular instances[10].

On Urbach's account, Bacon was not an infallibilist about the theories generated by his method[11]. Bacon's use of the term "certain" was, according to Urbach[12] , akin to the contemporary scientist's use of the term "verified" – and merely meant "well-confirmed". Generally, and ironically given Popper's determined attack on Bacon, Urbach argues that there is a considerable similarity between Bacon and Popper[13]. Like Popper, Bacon attacked conventionalism – in his day in astronomy – because conventionalism disallowed speculation about unobservables, and because conventionalism encourages ad hoc adjustments of theory[14]. So the primary defect of the method of anticipation lay, in Bacon's view, in its attachment to the immediately observable[15].

## 8.3 Computational Baconianism

Baconianism of an apparently unsophisticated and "standard" sort has re-surfaced in the form computer programs written by Herbert Simon and his colleagues at Carnegie-Mellon University[16].

---

[9]Urbach [1987] p. 35.
[10]Urbach [1987] p. 27.
[11]Urbach [1987] p. 38.
[12]Urbach [1987] p. 43.
[13]Urbach [1987] p. 49ff.
[14]Urbach [1987] p. 31.
[15]Urbach [1987] p. 33.
[16]See Simon et al. op. cit.

Simon's computational Baconianism is Bacon's *Novum Organum*, published in 1620, plus John Stuart Mill's Canons of Induction, plus cognitive science, implemented on the computer. Simon and his team have written half a dozen programs called *BACON.1*[17] through *BACON.6* along with a few more programs, like *Dalton* and *Glauber*[18]. Given facts these programs can extract scientific laws.

Told about the positions and periods of the planets even the most primitive version of Bacon – BACON.1 – can re-discover Kepler's Third Law of planetary motion, which tells us that

$$\frac{P^2}{d^3} = k$$

where $P$ is period of orbit of a planet, $d$ is its distance from the sun, and $k$ is some constant.

In Simon's account of his work, Mill is the source of one of the *heuristics* which control the simplest of Simon's suite of programs, namely Mill's Method of Agreement[19]. Contemporary cognitive science supplies the view that man is an information-processing organism and the idea that information-processing can and should be modelled by computer programs. Add Simon's idea that science is essentially problem-solving and you have the New Baconianism, according to which

$$\textit{Science} = \textit{Problem-Solving}$$

and

$$\textit{Problem-Solving} = \textit{Computation} + \textit{Heuristics}$$

---

[17] We refer to Simon's program as BACON.1, and the Prolog implementation described in this chapter as Bacon 1.0.

[18] Give the chemistry program Glauber data about how acids and alkalis react to form salts, and it can discover the concepts of "acid" and "alkali" and the law that they tend to react with one another. Surprisingly perhaps, Simon's programs happen to be rather simple. His team has written them in a language called PRISM which rides on top of - or is implemented in - the old near-functional programming language Lisp. Coded up in Prolog they typically take up just a few pages, as we shall see.

[19] Simon et al op.cit. p. 75. Actually, Simon calls the Method of Agreement the Rule of Agreement.

therefore

$$\text{Science} = \text{Computation} + \text{Heuristics}$$

Computational Baconianism is in one respect one big step forward for inductivism in that it projects a puzzle in the philosophy of science into the world of information technology where competing solutions of the puzzle may, in certain respects, be tested. It is also meant to be a scientific, and so testable, theory of science. But, as I shall argue, computational Baconianism also appears to take several steps backwards, and seems to discard, or at least ignore, most of the progress that philosophers of science usually think their subject has made this century.

## 8.4 Computational Baconianism as Science

The new Baconianism is apparently so different from anything approximating to contemporary orthodoxy in the philosophy of science that it is best, at least initially, to think of it as quite orthogonal to it. It is intended to be a *scientific theory* about scientific discovery, and not a *philosophical justification* of science as it is. The general idea is that

> ...the mechanisms of scientific discovery are not peculiar to that activity but can be subsumed as special cases of the general mechanisms of problem solving...[20]

Problem solving in science as elsewhere employs the general mechanism of *heuristic selective search.* Following the account of the problem solving process given by contemporary cognitive psychology, Simon views the human brain as an information processing system which solves a problem by constructing a space of possible solutions, the

[20]Simon et al op.cit. p. 5.

solution to the problem being found by a search through the solution space using heuristics[21].

According to Simon's hypothesis, scientific discovery is just problem solving of a special kind. Much discovery is *data-driven*, he holds, rather than *theory-driven*, as currently fashionable pictures of science suppose. Pure data-driven discovery can employ only *weak* heuristics, for the stronger the heuristics one uses in the problem-solving process, the greater the theoretical preconceptions with which one is viewing the data. The question is then:

> ...Can a program that contains only these selective heuristics to guide it actually discover significant scientific laws with only a modest amount of computing effort? If an affirmative answer can be given to this question, then we can claim to have driven the mystery out of these simple kinds of scientific discovery at least[22].

Simon's answer *is* affirmative, and the success of his Bacon programs constitutes a part of the evidence for his answer. Computational Baconianism is deflationary, proposing a largely data-driven theory of scientific discovery, the latter being viewed as discovery of empirical laws embedded in data.

Three questions arise.

First, what alternative is there to the standard view of Bacon?

Secondly, to what extent does computational Baconianism implement the "standard view"?

Thirdly, can computational Baconianism offer a satisfactory theory of scientific discovery, and so offer offer support to a Baconian view of science?

---

[21]There are weak heuristics and there are strong heuristics. Weak heuristics have general application, and include such strategies as hill-climbing - in which one short-sightedly evaluates progress towards the solution by comparing the relative quality of locally available solutions - and means-end analysis - in which one compares the current "solution" with some property of the desired solution. Strong heuristics are specific to particular domains, and sacrifice general applicability for efficiency in those domains.

[22]Simon et al op.cit. p.33.

In this chapter we describe how simple Baconian programs discover laws of nature. Simon's Bacon sequence of programs succeed in extracting simple quantitative scientific laws from data. We prototype a version of the simplest program in the sequence – BACON.1 – in the logic programming language Prolog. Bacon 1.0, our prototype, illustrates the essence of Simon's New Baconianism, and we follow it up with a philosophical commentary.

# Section 2

## 8.5 Bacon 1.0: A Prolog Prototype[23]

Bacon 1.0 requires two lists of numerical values of two dependent variables X and Y of data. Call the lists of values of X and Y the X_List and Y_List respectively. The X_List $[X_1, X_2, ..., X_k]$ and Y_List $[Y_1, Y_2, ..., Y_k]$ have equal lengths $k$. $X_i$ and $Y_i$ – for $0 \leq i \leq k$ – are the correlated values of X and Y for the $i$th observation. X and Y may be called the *dependent* variables, for the *independent* variable $i$.

## 8.6 Heuristics[24]

Bacon 1.0 seeks a term of the form $X_i^m Y_i^n$, which is constant for all the $k$ values of the independent variable $i$. To achieve this result, it employs the following heuristics. It first checks to see:

(a) *if the X- and Y-lists increase and decrease together, that is, if they are* increasing[25] – and

[23] A version of Bacon 1.0 running on the Apple Macintosh and entitled MacBACON 1.00 is available from Episteme Ltd, 23 Five Mile Drive, Oxford.

[24] BACON.1 also searches for linear relationships between X and Y, and for constant values of X and Y. It is relatively simple to extend the heuristics presented here to cover this case, and MacBACON 1.00 does so.

[25] INCREASING and DECREASING are not a good choice of term. See Simon et al op.cit. p.77. In Simon's terms the lists [1,3,2,5,4] and [5,7,6,10,9] are increasing. A better pair of terms might have been *similarly_directed* and *oppositely_directed.*

(b) *if the X- and Y-lists are such that as one increases, the other decreases, that is, if they are* decreasing.

If the X- and Y-lists increase together, the program generates the list formed by considering the *ratios* $X_i/Y_i$, that is $X_i^1 Y_i^{-1}$. If, on the other hand, the X- and Y-lists are such that as one increases, the other decreases, the program generates the list formed by considering the products of $X_i Y_i$, that is $X_i^1 Y_i^1$. If neither of these alternatives obtains then Bacon 1.0 fails to extract a law from the data[26].

---

[26]What do we mean by *the X- and Y-lists increase together*? We can take either a strict or a lazy view. Simon's BACON program takes the strict view. This might be specified as follows. Given two lists X and Y of equal length L, we specify the partial function:

strictly_increasing: **seq of** $\mathbb{R}$ × **seq of** $\mathbb{R}$ → **B**
strictly_increasing(X,Y) $\triangleq \forall i, j \in \{1..L\} \bullet (X[i] < X[j] \Leftrightarrow Y[i] < Y[j])$

Bacon 1.0 takes the lazy view, which might be specified as follows:

lazily_increasing: **seq of** $\mathbb{R}$ × **seq of** $\mathbb{R}$ → **B**
lazily_increasing(X,Y) $\triangleq$
**if** L ≤ 1 **then** *true*
**else** ((X[1] < X[2] ∧ Y[1] < Y[2]) ∨ (X[2] < X[1] ∧ Y[2] < Y[1]))
∧
lazily_increasing(tail(X),tail(Y))

This lazy specification is captured in the Bacon 1.0 code for **increasing/2**.

```
increasing([],[]).
increasing([_],[_]).
increasing([X1,X2 | T1], [Y1, Y2 | T2]) :-
    ((X1 > X2, Y1 > Y2)
    ;
    (X1 < X2, Y1 < Y2)),
    increasing([X2 | T1], [Y2 | T2]).
```

Given a pair of lists which are either both the empty list, or are both lists with a single element, then both lists are *increasing*. This is the content of the first two clauses. Otherwise both lists have more than two elements. Such a pair of lists will be *increasing* if the first two elements are both increasing or both decreasing, and both tails of the two lists are *increasing*. Similar code handles **decreasing/2**.

How do the procedures consequent on (a) and (b) work in practice? Suppose that the first of these alternatives obtains – the X- and Y-lists are *increasing* – so that the program has generated the new list $[X_1^1 Y_1^{-1}, X_2^1 Y_2^{-1}, \ldots, X_k^1 Y_k^{-1}]$. Given the X- and Y-lists, we can represent this new list by the term representation [1, -1]. (Generally, the list [m, n] represents the term $X^m Y^n$.)

Now we have three lists, the original X-list and Y-list together with a new list of values formed from the original X- and Y-lists. Bacon 1.0 continues this process, comparing the values of lists produced by the generated terms, until the generated list is constant list[27].

At each step the program decides whether or not to multiply or divide the new list by X or Y, by comparing the latest generated list with the X- and Y-lists according to the heuristics (a) and (b). To avoid generating lists that have *already been generated* the program has to keep track of which terms have already been formed. So we have a third heuristic:

(c) *Bacon 1.0 avoids forming a previously formed term even if the heuristics of (a) or (b) suggest that it should.*

---

Notice that the strict and the lazy specifications are not equivalent in general. For example,
X = [1, 3, 2, 4] and Y = [1, 4, 2, 3]
are increasing in the lazy sense. Corresponding successive pairs in the X and Y lists increase and decrease correspondingly. But X and Y are not increasing in the strict sense because X[2] < X[4] while it is not the case that Y[2] < Y[4].

However, if the X and Y data lists are related by a law of the form $X^m Y^n$ = *constant*, then such a situation cannot arise. If, for example, the X and Y lists are increasing, then n is negative. Therefore if X[2] < X[4], it follows that Y[2] < Y[4]. Therefore, if there is a law of the form $X^m Y^n$ = *constant*, then the strict and lazy specifications are equivalent.

[27]This is Simon's application of the Mill's Method of Agreement. (See Simon op. cit. p. 75.) Note that Mill's Method of Agreement (Mill [1961] p. 255) - his First Canon - deals with discovering causal laws rather than purely mathematical relationships. It is formulated as follows: If two or more instances of the phenomena under investigation have only one circumstance in common, the circumstance in which alone all the instances agree is the cause (or effect) of the given phenomena. Simon reads this as: if the values of a pair of independent variables describing the phenomena (values of the dependent variable) are described by a simple mathematical relationship, then this mathematical relationship in which all the instances agree is the law describing the given phenomena.

Finally, we are not interested in the law that $X^0Y^0$ is a constant, since it is trivially true. So we have the additional heuristic[28]:

(d) *Avoid the law* $X^0Y^0$ = *constant.*

The essential code for Bacon 1.0 is listed in Appendix 1.

## 8.7 Kepler's Third Law Traced

How does Bacon 1.0 fare with Simon's[29] artificial data lists for Kepler's third law? The data are

| Planet (i) | period (X) | distance (Y) |
|---|---|---|
| 1 | 1.0 | 1.0 |
| 2 | 4.0 | 8.0 |
| 3 | 9.0 | 27.0 |

With this the following query

|?- go_bacon([1.0,4.0,9.0],[1.0,8.0,27.0]).

writes a trace to the screen exhibited in the following photo-log[30]

```
|?- go_bacon([1.0,4.0,9.0],[1.0,8.0,27.0]).

    Term is  [1,-1]  [[1,-1]]                          [1, 0.5, 0.333333]
    Term is  [2,-1]  [[2,-1],[1,-1]]                   [1, 2, 3]
    Term is  [2,-2]  [[2,-2],[2,-1],[1,-1]]            [1, 0.25, 0.111111]
    Term is  [3,-2]  [[3,-2],[2,-2],[2,-1],[1,-1]]     [1, 1, 0.999999]

Law is: X^3 .Y^-2 = 0.999999
```

---

[28]In the terminology of heuristic search theory, the heuristics, the X_List, the Y_List, and the Term_list – the list of terms of the form [m,n] – constitute the *database*. The heuristics (a) and (b) make up the *control strategy*. The machinery for generating new members of Term_list constitutes the *production rules*. The control strategy is *systematic* in that heuristic (c) protects against inefficient backtracking.

[29]Simon et al. op. cit. p. 66ff.

[30]A photo-log is a record of the output delivered to the screen during the running of a program. Input, in the form of keyboard work, appears in the photo-log exactly as it appears on the screen. In the photo-log of Prolog sessions, text input from the keyboard has been edited to appear italicised.

The three columns in the output display the current term [m,n] means $X^mY^n$; the list of terms examined so far; and finally the values of $X_i^mY_i^n$ for $1 \leq i \leq 3$.

So Bacon 1.0 "discovers" Kepler's third law. The Term [3,-2] means $P^3/d^2$. The constant value is represented by the list [1, 1, 0.999999].

We can display the query:

```
|?- go_bacon([1.0,4.0,9.0],[1.0,8.0,27.0]).
```

with the positions of the X- and Y-lists interchanged in tree form, as shown in Figure 1.

The order of the lists in the go_bacon query go_bacon doesn't matter, though the efficiency of the search may be impaired, as it is in the following query[31]:

```
|?- go_bacon([1.0,8.0,27.0],[1.0,4.0,9.0]).

    Term is  [1,-1]  [[1,-1]]                       [1,2,3]
    Term is  [0,-1]  [[0,-1],[1,-1]]                [1,0.25,0.111111]
    Term is  [0,0]   [[0,0],[0,-1],[1,-1]]          [1,1,1]
    Term is  [1,-2]  [[1,-2],[1,-1]]                [1,0.5,0.333333]
    Term is  [2,-2]  [[2,-2],[1,-2],[1,-1]]         [1,4,8.99999]
    Term is  [2,-3]  [[2,-3],[2,-2],[1,-2],[1,-1]]  [1,1,0.999999]
```

Law is: X^2 .Y^-3 = 0.999999 Again Bacon 1.0 discovers Kepler's Third Law. The Term [2,-3] means $d^2/P^3$.

---

[31]Label the six lines of the trace (1)-(6). The execution of this query exhibits some interesting backtracking at line (3). At line (1) initialization occurs. X_List and Y_List are similarly directed so the initial term is [1,-1] chosen, signifying X/Y. The Result_list for [1,-1] is [1,2,3]. At line (2) the X_List and the Result_list are found to be similarly directed, generating division by X, leading to the term [0,-1]. Line (3) is the interesting one. At line (3) the new Result_list [1,0.25,0.111111] is first found to be oppositely directed with respect to the X_List, but multiplication by X would lead back to the initial [1,-1], and secondly found to be oppositely directed with respect to the Y_List, leading to the term [0,0]. We refuse to terminate on this term, according to heuristic (d), so the search backtracks to the initial Term list [[1,-1]]. Since the original Result_list [1,2,3] and the Y_List are similarly directed, the term tested at line (4) is [1,-2] obtained by dividing by Y. From then on the search is straightforward.

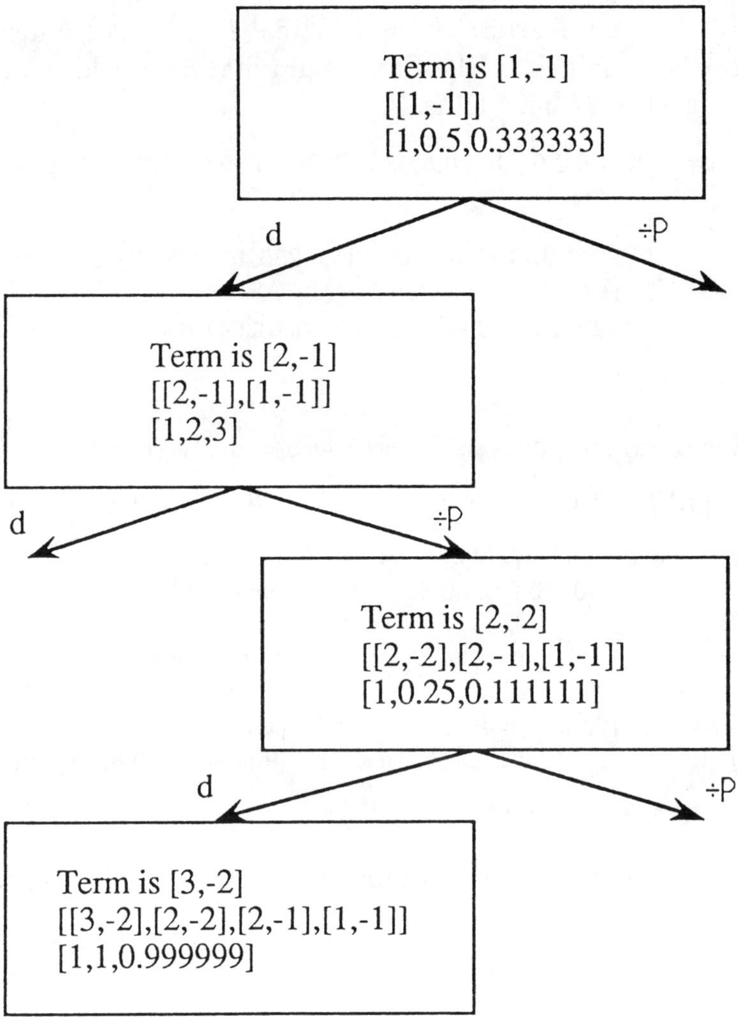
Term is [1,-1]
[[1,-1]]
[1,0.5,0.333333]
d
÷P
Term is [2,-1]
[[2,-1],[1,-1]]
[1,2,3]
d
÷P
Term is [2,-2]
[[2,-2],[2,-1],[1,-1]]
[1,0.25,0.111111]
d
÷P
Term is [3,-2]
[[3,-2],[2,-2],[2,-1],[1,-1]]
[1,1,0.999999]

**Figure 1**

## 8.8 Galileo's Law Traced

The photo-logs for Galileo's idealized[32] data show that the order of the lists doesn't matter. First for time $t$ against distance travelled $d$, for a body falling in the Earth's gravitational field, and then Distance_travelled against Time.

```
|?- go_bacon([0.1,0.2,0.3,0.4,0.5,0.6], [0.098,0.392,0.882,1.568,2.450,3.528]).

  Term is  [1,-1]    [[1,-1]]
           [1.02041,0.510203,0.340136,0.255102,0.204082,0.170068]
  Term is  [2,-1]    [[2,-1], [1,-1]]
           [0.102041,0.102041,0.102041,0.102041,0.102041,0.102041]

Law is: X^2 .Y^-1 = 0.102041
```

For this data, the constant list is the obviously rigged

[0.102041,0.102041,0.102041,0.102041,0.102041,0.102041]

```
 |?- go_bacon(  [0.098,0.392,0.882,1.568,2.450,3.528],
                [0.1,0.2,0.3,0.4,0.5,0.6],Term).

    Term is  [1,-1]  [[1,-1]]                [0.98,1.96,2.94,3.92,4.9,5.88]
    Term is  [0,-1]  [[0,-1],[1,-1]]         [10,5,3.33333,2.5,2,1.66667]
    Term is  [0,0]   [[0,0],[0,-1],[1,-1]]   [1,1,1,1,1,1]
    Term is  [1,-2]  [[1,-2],[1,-1]]         [9.8,9.8,9.8,9.8,9.8,9.8]

Law is: X^1 .Y^-2 = 9.8
```

The law is $t^2/d$ = constant or, equivalently, $d/t^2$ = constant.

## 8.9 Termination

Does Bacon 1.0 terminate and terminate correctly whenever it is given lists X and Y satisfying the law $X^m Y^n = constant$? The answer is that it does, and with $m$ and $n$ in their smallest values.

We can always write the law $X^m Y^n = constant$ in a form in which $m > 0$. This is the form of the law which Bacon 1.0 finds. We therefore *assume* that $m > 0$.

[32]Simon et al op.cit. p. 83.

Consider first the case in which $n > 0$.

**$n > 0$**

Given the lists X and Y, Bacon 1.0 initializes a list of terms called Term_list. When $n > 0$, X and Y are *decreasing*, so Bacon 1.0 initializes

$$\text{Term\_list} = [[1,1]]$$

*We show that the list of terms which Bacon 1.0 sets about building consists of terms of the form* $[m', n']$*, where* $m \geq m' \geq 0$ *and* $n \geq n' \geq 0$.[33]

For each of the terms $[m', n']$, Bacon calculates a Result_list R of the form

$$[X_1^{m'} Y_1^{-n'}, X_2^{m'} Y_2^{-n'}, \ldots, X_k^{m'} Y_k^{-n'}]$$

Two of the following four conditions

(i) decreasing(X,R)

(ii) increasing(X,R)

(iii) increasing(Y,R)

(iv) decreasing(Y,R)

will always apply, unless either $m' = 0$ or $n' = 0$.

If *both* $m' = 0$ *and* $n' = 0$, the term is rejected in accordance with heuristic (d). If $m'/n' = m/n$, then Bacon 1.0 terminates correctly, with $m' = m$ and $n' = n$ (that is, with $m'$ and $n'$ in their lowest forms).

So there are two other possibilities.

**$n > 0 - m'/n' > m/n$**

Let R be the Result_list generated by the current term $[m', n']$. In this case (ii) and (iv) apply. Case (ii) decreases the m-value and generates $[m' - 1, n']$. Case (iv) increases the n-value, and generates $[m', n' + 1]$, from which it follows that if $m'/n' > m/n$, then the n-value cannot be rendered negative.

---

[33] In other words, $\forall m', n' \bullet ([m', n'] \in \mathit{Term\_list} \Rightarrow 0 \leq m' \leq m \wedge 0 \leq n' \leq n)$

(a) If $m'/n' > m/n$ the algorithm of Bacon 1.0 first applies case (ii) – increasing(X,R) – and generates the term $[m' - 1, n']$.

If the m-value $m'$ were the first of either the m-value or the n-value of the term $[m', n']$ to exceed the desired results $m$ and $n$ respectively, case (ii) would immediately return $m'$ to $m$. So that $m'$ is contained with an envelope whose upper limit is $m + 1$.

So can the process of applying case (ii) generate a term whose m-value is negative?

Assume [*] that the first occurrence of a *negative* value for either $m'$ or $n'$ in the Term_list occurs for $m'$. (Remember that we are assuming that both $m$ and $n$ are greater than zero.) The previous term generated must therefore have been of the form $[0, n']$. But this term – $[0, n']$ – must have generated a Result_list $R'$, for which case (iii) applies – increasing(Y,$R'$) – because $n'$ is assumed to be greater than zero. A term of the form $[0, n']$ would therefore have successively generated terms of descending n-value finally reaching [0, 0]. At that point the heuristic would have rejected the original choice of case (ii). So case (ii) cannot make the m-value of the term negative, unless the n-value is already negative, but this contradicts our assumption [*].

Therefore $m'$ is contained with an envelope whose lower limit is zero, *unless $n'$ is already negative.*

(b) Transferring to case (iv) – decreasing(Y,R) – generates a term of the form $[m', n' + 1]$. His n-value – $n' + 1$ – cannot be the first of either the m-value and the n-value to exceed the desired result – in this case $n$ – for the following reason. If in the term $[m', n' + 1]$, $n' + 1$ were to exceed $n$ before the m-value $m'$ of the term exceeded $m$, it would first have to pass through $[m', n]$ before reaching $[m', n' + 1]$. But in that case, the heuristics of Bacon 1.0 would apply case (i) – decreasing(X,R) – and would generate a sequence of terms $[m', n]$, $[m' + 1, n]$, $[m' + 2, n]$, ..., $[m, n]$, which would terminate correctly.

A similar argument to the above shows that $n'$ is contained with an envelope whose lower limit is zero, *unless $m'$ is already negative.* Since both $m' = n = 1$ initially, neither can become negative. Both are contained in an envelope whose lower bound is $m' = n' = 0$.

**$n > o - m'/n' < m/n$**
An argument similar to that for the case $m'/n' > m/n$ shows that the n-value of a term in Term_list cannot be negative, and that the m-value of the term cannot be the first of either the m-value and the n-value to exceed the desired result, namely $m$.

If $n > 0$, then the four cases (i)-(iv) constrain the search in the way shown in Figure 2.

**$n < o$**
A dual argument shows that Term_list can never contain a term $[m', n']$ in which $m' < 0$ or $n' > 0$, and in which either $m' > m$ or $n' < n$.

Taking these cases together, one sees that the number of possible Term_lists generated by Bacon 1.0 from given data satisfying a law of the form $X^m Y^n = constant$, is finite. There can be no loops in the execution of the program because Term_list cannot contain repeated terms.

Therefore if there is a law of the form $X^m Y^n = constant$, Bacon 1.0 terminates with the correct answer.

# Section 3

## 8.10 Is Computational Baconianism Baconian?

Bacon 1.0 is clearly conjectural. Like Urbach's Bacon, it exhibits a *hypothetico-inductive* methodology. It is a conjecturing and refuting machine. Its theories – equations of the form $X^m Y^n = constant$ – are generated, and then tested against (falsified by) the data. What makes the process appear to be naively inductive is the manner in which the program decides which terms to conjecture. For these are specified by the program's heuristics, for these terms are partly determined by the data, the X- and Y-lists.

A Bacon 1.0 style program has no way of deciding which data to search. But Urbach argues that Bacon never thought that it was[34]. It

[34]See Urbach [1987] pp. 155-6.

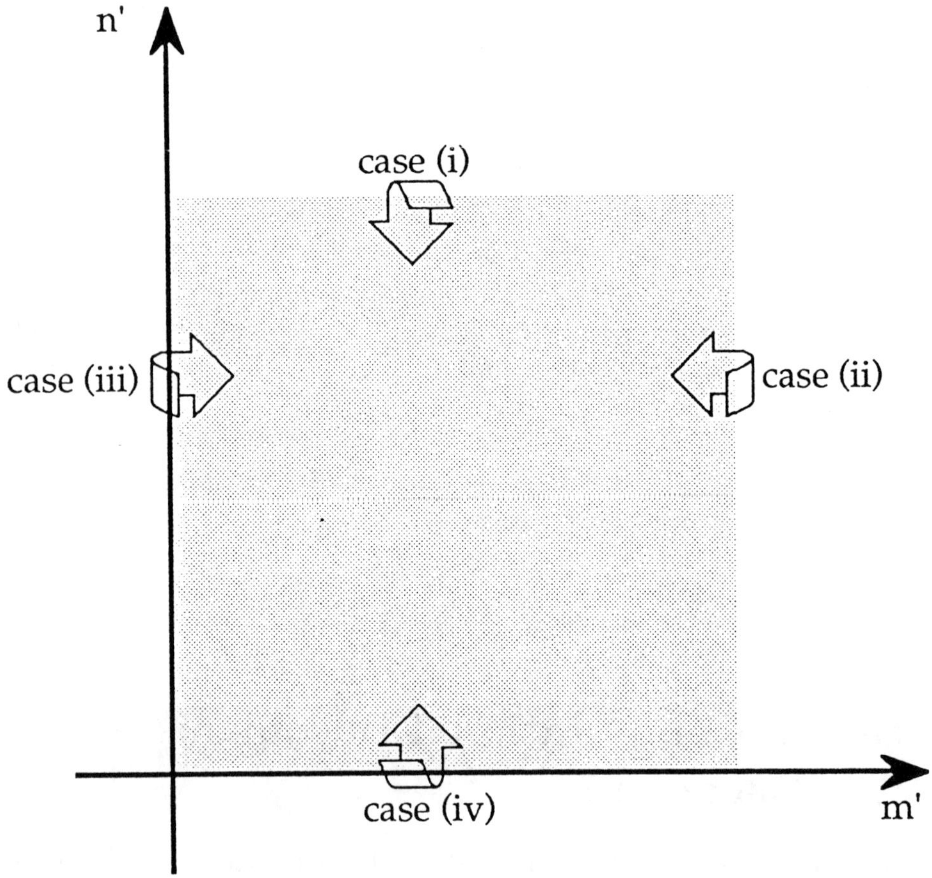


**Figure 2**

is at least arguable that there is nothing un-Baconian about Bacon 1.0 in this respect. In addition, and like Bacon the philosopher, Bacon 1.0 does not presume perfectly accurate data – notice the $\Delta$ error in the requirement for **is_constant/3**. In these senses, the Book of Nature isn't open to the program.

So far so Bacon-like.

The most obvious objection to Simon's programs as implementations of a specifically Baconian methodology is that they embody anticipations – theories which the program tenaciously adheres to. Anticipations are built into the heuristics of the programs. Bacon 1.0 has a theory built into it – that a law of the form $X^m Y^n = constant$ holds. Why should we expect laws of precisely this form, out of the infinity of possibilities? In this anticipatory respect Simon's Bacon – BACON.1 through BACON.6 – programs are quite un-Baconian, unless they are used as a mere assistant by a Baconian empirical scientist who can exercise judgement about when to employ them.

## 8.11 The Larger Issue

Does computational Baconianism supply us with an adequate conception of scientific discovery? I doubt if many people would think so. But why?

*Perhaps it is because a Bacon style program seems too stupid to do any real learning.*

In Bacon 1.0 we build the heuristics which determine the direction in which the search-space is searched into the clause for **search_for_term/1**. The clause which terminates the search and determines the discovered law is for **terminate/4**. It would not be too difficult to write an enhanced version of the program which would **assert** further clauses for these functors as it made discoveries. Such a program would modify itself, and would, in a sense, learn.

It might then be objected that its power to learn would be similarly circumscribed by the anticipations of the programmer. But if the enhanced Bacon 1.0 were ruled out as a genuine Baconian discovery program on that score, then so would any program. And the objection would be not that Simon's programs were un-Baconian, but that any algorithmic version of scientific discovery, however complex, however much its behaviour exceeded our capacity to predict it, would be too.

*Perhaps it is because the Bacon program leaves out too much of the scientist's mental life.*

To this, Simon has an answer. To the claim that mental imagery plays an essential role in scientific discovery Simon says:

> "mental imagery as a mode of thought can be explained in the same terms as any other kind of problem-solving representation: as a set of symbol structures on which certain operations can be performed to make explicit some of the information that is initially implicit."[35]

If mental imagery plays a role in scientific discovery, it must function as just another form of expertise, or *effective heuristic.* There is therefore merely a variety of representations:

> "[b]y a representation we mean a scheme for holding information in a memory, combined with processes for operating on that information to access it, alter it, or draw inferences from it..."[36]

> "...two representations are equivalent only to the extent that the same information can be stored in them and extracted from them with the same ease ... the essential role of the basic operations in defining a representation..."[37]

But different representations are

> "...are not computationally equivalent. Inferences that are drawn easily with the use of operators associated with one representation may be drawn only with difficulty, or not at all, if the operators associated with another representation are used."[38]

[35]Simon et al op.cit. p.336.
[36]Simon et al op.cit. p.315.
[37]Simon et al op.cit. p.316.
[38]Simon et al op.cit. p.322.

> "...with every field of expertise is associated one or more specific representations, with which are associated, in turn numerous operators that permit the expert to extract from situations large amounts of information that is not available to novices..."[39]

These, especially the last, are *scientific* hypotheses They fit in with the information-processing model of man. And they have the miniscule evidence in their favour afforded by the success of the Bacon programs, and their successors. I say "miniscule" for the reasons that the Dreyfuses[40] have advanced. The Bacon programs haven't done anything hard yet, like real science. One day they might, though the unsuccess of AI as technology doesn't give one much confidence that they will. Nevertheless we must keep an open mind.

*Perhaps it is because the Bacon programs caricature the history of science. If science were this easy, Kepler and his predecessors must have been pretty stupid.*

If scientific activity were as computational Baconianism describes it, it would be too easy and we should have trouble explaining why science took so long to develop. Bacon 1.0 assumes, as Kepler did, that there are simple equational laws which describe the planetary motions. In Sir Francis Bacon's terms, both the program and Kepler *anticipated* nature. In Kepler's case, the anticipation sprang from his neo-Platonism. One has to say that in the case of Bacon 1.0 the anticipation springs from the programmer's knowledge of the law he expects the program to discover.

Why choose the particular variables of period $P$ and distance $d$ from the sun? As with the *frame problem*, the problem is one of *relevance*. We should say, I think, that the variables we choose to put into our equations are a product of our particular way of representing the world, and that this is a product of our language, of our history, and of the way we as organisms and agents interact with the world.

Perhaps this is part of what makes computational Baconianism seem so implausible, so *thin*, as a theory of scientific discovery its over-abstraction from real scientific activity – *science-in-the-world*, as one might say. The implausibility of computational Baconianism stems, so

[39] Simon et al op.cit. p.323.

[40] See Dreyfus [1986].

I want to say, from a failure to see that science is one of the products of the encounter between human beings who have grown up in the world, and the world itself.

## Appendix 1

```
go_bacon(X_List,Y_List) :-
initialise(Initial_term,X_List,Y_List),
search_for_term([Initial_term]).

initialise(Initial_term,X_List,Y_List) :-
update(x_list(X_List)),
update(y_list(Y_List)),
!,
( decreasing(X_List,Y_List),
Initial_term = [1,1],
!
;
increasing(X_List,Y_List),
Initial_term = [1,-1]
).

update(X) :-
functor(X,Functor,Arity),
abolish(Functor,Arity), !,
asserta(X).

search_for_term([]) :- !, fail.

search_for_term(Term_list) :-
x_list(X_List),
y_list(Y_List),
Term_list = [ [P1, P2] | Tail_term],
result_list([P1,P2],X_List,Y_List,Result_list),
!,
trace_calculation(P1,P2,Term_list,Result_list),
!,
(terminate_constant([P1,P2],Result_list)
```

```
;
x_list(X_List),
(decreasing(X_List,Result_list),
P is P1 + 1
;
increasing(X_List,Result_list),
P is P1 - 1 ),
not(member([P,P2],Term_list)),
search_for_term([ [P,P2] | Term_list])
;
y_list(Y_List),
(increasing(Y_List,Result_list),
P is P2 - 1
;
decreasing(Y_List,Result_list),
P is P2 + 1 ),
not(member([P1,P], Term_list)),
search_for_term([ [P1,P] | Term_list]) )
;
search_for_term(Tail_term).

trace_calculation(P1,P2,Term_list,Result_list) :-
write('Term is '),
write([P1,P2]),
tab(10),
write(Term_list),
tab(10),
write(Result_list),nl.

/* */

decreasing([],[]).
decreasing([_],[_]).
decreasing([X1,X2 | T1], [Y1, Y2 | T2]) :-
((X1 > X2, Y1 < Y2)
;
(X1 < X2, Y1 > Y2) ),
decreasing([X2 | T1], [Y2 | T2]).
```

```
increasing([],[]).
increasing([_],[_]).
increasing([X1,X2 | T1], [Y1, Y2 | T2]) :-
((X1 > X2, Y1 > Y2)
;
(X1 < X2, Y1 < Y2) ),
increasing([X2 | T1], [Y2 | T2]).

member(X, List) :-
List = [X|_] ; (List = [_|Tail], member(X,Tail)).

/* */

is_constant(List,Mean,Error) :-
(List = []
;
List = [_] ),!
;
mean(List,Mean),
Delta is Mean * Error,
inside_error(List,Mean,Delta).

sum(List, Sum) :-
List = [], Sum = 0
;
List = [Head | Tail],
sum(Tail, Sum_1),
Sum is Head + Sum_1.

mean(List,Mean) :-
sum(List,Sum),
length(List,Length),
Mean is Sum / Length.

inside_error(List,Mean,Error) :-
mean(List,Mean),!,
within_error(List,Mean,Error).

within_error(List,Mean,Error) :-
```

```
List = []
;
(List = [Head | Tail],
Difference is Mean - Head,
absolute_value(Difference, Difference_1),
Difference_1 < Error,!,
within_error(Tail,Mean,Error)).

absolute_value(X,Positive) :-
X =< 0, Positive = - X,!
;
X = Positive.result_list(_,[],[],[]).

result_list([P1,P2],[H1 | T1], [H2 | T2],[H | T]) :-
power(H1,P1,X),
power(H2,P2,Y), H is X * Y,
result_list([P1,P2],T1,T2,T).

power(X,N,Y) :- /* y = x to the n */

(N = 0, Y = 1.0)
;
(N > 0, N1 is N - 1, power(X,N1,Y1), Y is X * Y1)
;
(N < 0, N1 is N + 1, power(X,N1,Y1), Y is Y1 / X).

terminate_constant([P1,P2],Result_list) :-
( not(P1 = 0)
;
not(P2 = 0) ),
is_constant(Result_list,Constant,0.01), nl,
write('Law is: X^'),write(P1),write(' .Y^'),write(P2),
write(' = '),write(Constant),nl.
```

## References

Dreyfus, H.L., and S.E., [1986]: *Mind over Machine*, Basil Blackwell. Oxford.

Langley P, Simon H.A., Bradshaw G.L. and Zytkow J.M., [1987]: *Scientific Discovery - Computational Explorations of the Creative Process.* The MIT Press. Cambridge, Ma.

Mill, J.S., [1961]: *Collected Works Vol. XXX*, Routledge.

Popper, K.R., [1968]: *Conjectures and Refutations*, Harper Torchbooks. New York.

Popper, K.R., [1972]: *Objective Knowledge*, The Clarendon Press. Oxford.

Urbach, P., [1982]: "Francis Bacon as a Precursor to Popper", *The British Journal for the Philosophy of Science*, 33, pp. 113-132.

Urbach, P., [1987]: *Francis Bacon's Philosophy of Science*, Open Court. La Salle, Illinois.

# Chapter 9

# Causality from Probability

Peter Spirtes, Clark Glymour and Richard Scheines[1]
*Department of Philosophy, Carnegie-Mellon University, Pittsburgh, USA*

## 9.1 Introduction

### 9.1.1 Uses of data analysis for policy generally involve causal inference

Data analysis that merely fits an empirical covariance matrix or that finds the best least squares linear estimator of a variable is not of itself a reliable guide to judgements about policy, which inevitably involve causal conclusions. The policy implications of empirical data can be completely reversed by alternative hypotheses about the causal relations of variables, and the estimates of a particular causal influence can be radically altered by changes in the assumptions made about other dependencies.[2] For these reasons, one of the common aims of empirical research in the social sciences is to determine the causal relations among a set of variables, and to estimate the relative importance of various

[1]The research in this chapter was supported by the Office of Naval Research under Contract number N00114-88-K-0194. The second author was supported by a fellowship from the Simon Guggenheim Memorial Foundation.

[2]See our discussion of the causal relations between foreign capital on political repression in [Glymour 1987].

causal factors. Even where that aim is not acknowledged it is often tacit. A question of first importance about empirical social science is therefore: how are causal relations amoung variables to be discovered?

### 9.1.2 The difficulty of the discovery problem

The difficulty of this question is apparent when one considers the number of possible causal models for a given set of variables. If the causal dependence of one variable on another is represented by a directed edge from a vertex representing the causal variable to a vertex representing the effect variable, then the number of possible causal structures on $n$ variables is the number of directed graphs with $n$ vertices, or $4^{\binom{n}{2}}$. If causal cycles are forbidden, then the number of possible causal structures on $n$ variables is the number of acyclic directed graphs on $n$ variables. For 12 variables the number of directed graphs is approximately $5.4 \times 10^{39}$ and the number of acyclic graphs is 521, 939, 651, 343, 829, 405, 020, 504, 063 [Hilary 1973]. Even when the time order of the variables is known , so that causal hypotheses in which later variables cause earlier variables can be eliminated, the number of alternatives remaining is generally very large: for 12 variables it is $7.4 \times 10^{19}$.

The social scientist who addresses a problem area where causal questions are of concern is therefore faced with an extremely difficult discovery problem, for which there are only three avenues of solution: (i) use experimental controls to eliminate most of the alternative causal structures; (ii) introduce prior knowledge to restrict the space of alternatives; and (iii) use features of the sample data to restrict the space of alternatives.

Experimental procedures for addressing social questions are much to be desired, but they are very expensive and often infeasible. Where quasi-experiments are used that control some variables but not others, the number of alternative causal structures possible *a priori* may remain very large. Generating the set of admissible causal structures from "substantive theory" is recommended routinely in methodology texts.[3] In practice publications in the social science literature usually restrict the number of alternatives considered to a very few, and the restrictions

[3]See, for example, [Joreskog 1984], [Duncan 1975].

are often justified by citing prior literature or by appealing to very broad theoretical frameworks. It is anybody's guess, however, whether such appeals constitute a reliable discovery procedure. It seems at least likely that appeals to theory introduce bias and often exclude the true causal relations amoung the variables of interest. What about the third avenue?

### 9.1.3 Causal inference from statistical samples

Sample data are routinely used in systematic ways for parameter estimation in a parameterized family of probability distributions, but are more rarely used explicitly or systematically to infer causal structure. To the contrary, methodologists routinely warn against such inferences. The common slogan "correlation does not imply causality" is generally given as a caveat against trying to infer anything about causal dependencies from statistical data. Methodologists routinely warn that "substantive knowledge", not sample data, should determine the causal structure of a model. Procedures that use the sample data are denounced as "data mining" or "ransacking". It would be difficult to find a textbook on statistical methodology for the social sciences that does not include these warnings.[4]

For all the ferocity of the denunciation of sample-based causal inference, it is hard to find any sober analysis that justifies the conviction that reliable inference of this kind is impossible. There are worst-case arguments that point out the unreliability of data-based inference if the sample size is small compared to the number of variables, but these objections are readily avoided by appropriate sampling. There is the wide experience of social scientists and psychologists with a variety of "exploratory" factor analysis programs, which many people hold to be quite unreliable. But factor analytic programs involve very specific assumptions that are not in the least necessary in any possible procedure for inferring causal structure from sample data. Common factor analysis programs assume, for example, that the functional dependencies are linear, that measured variables have no direct effects on one another, and that measured variables never have effects on latent variables or

[4] See [Loehlin 1987], or [James et. al., 1982].

factors [Loehlin 1987]. Each of these assumptions may be false in a domain; none of them are essential to the idea of inferring restrictions on causal structure from sample data.

Is the complaint against sample-based causal inference simply an unfounded prejudice? If so, the best way to show as much is to provide reliable procedures for using sample data to usefully narrow the class of causal structure that are *a priori* possible for the data, and to prove that the procedures are reliable.

## 9.2 The Claims

Subject to simple and plausible principles connecting causal dependencies with statistical dependencies we will describe automatic procedures that:

(1a) if given data for a number of random variables generated by sampling from a multinormal distribution determined by unknown causal dependencies among those variables and by an unknown probability distribution on the exogenous variables, will find a "small" set of alternative causal structures;

(1b) if given the population covariances will output a set of causal structures that with probability one includes the true structure;

(1c) are reliable enough to be useful on samples of realistic size;

(2) provide a reliable statistical condition sufficient to warrant the conclusion that the statistical dependencies amoung a set of jointly distributed random variables are due in part to a random variable not contained in the set; in other words, find a reliable sufficient condition for the introduction of "latent" variables;

(3) given a causal structure, will generate all the alternative causal structures that are "statistically equivalent" to the given structure.

The procedures we will describe are partially implemented in the TETRAD II program now being developed and tested at Carnegie Mellon University.

These claims are odd in that they mix causal language with talk of probability distributions. That mixture nonetheless agrees with a great deal of informal usage in biostatistics, quantitative social science and in applied statistics generally. In this chapter we will provide a precise sense of causal dependency which gives a computable necessary and sufficient condition for two variables to be "directly" causally connected. This principle can be used to form an undirected graph of causal dependencies from sample data. A second computable general principle connecting causality and statistical dependency suffices to restrict the class of possible orientations of the edges of the undirected graph; i.e., gives a collection of directed graphs of causal relations from the sample data. We will describe a procedure that uses these principles and that answers (1a). We will state a theorem that for linear models ensures that the procedure satisfies the requirement of (1b), and we will describe some runs on simulated data that address (1c). Using a third principle – which is provably consistent with the first two – connecting causal structures and statistical dependencies, we will report but not prove results that give solutions to (2) and (3).

## 9.3 Causality and Statistical Dependence

Many writers [Suppes 1970], [Skyrms 1980] have connected causality with statistical dependence. We suppose the following principle:

> Let V be a set of random variables with a joint probability distribution. We say that variables x, y ∈ V are **directly causally dependent** if and only if there is a causal dependency between x, y (either the value of $x$ influences the value of y or the value of y influences the value of x or the value of some third variable not in V influences the values of both x and y) that does not involve any other variable in V.
>
> **Principle I:** For all x, y in V, x and y are directly causally dependent if and only if for every subset S of V not containing x or y, x and y are not statistically independent conditional on S. [Spirtes 1989b].

Principle I permits a procedure that generates from sample data an undirected graph whose edges represent direct causal dependencies. One needs to test for each of the conditional statistical dependencies that are possible *a priori*, and then apply the principle to build the graph. We have implemented this procedure in the TETRAD II program. The program tests for vanishing partial correlations of all orders.

To orient the undirected graph constructed with Principle I, a further principle is needed. Consider the following pictures of possible causal dependencies amoung three variables.

| | |
|---|---|
| $A \rightarrow B \rightarrow C$ | $A \leftarrow B \leftarrow C$ |
| $A \leftarrow B \rightarrow C$ | $A \rightarrow B \leftarrow C$ |

We say that **B is directly causally dependent on A** provided A and B are causally dependent and the direction of causal influence is from A to B.

If A and B are directly causally dependent and B and C are directly causally dependent, but A and C are not, then the relations between A and B, on the one hand, and B and C on the other hand, may produce an indirect causal dependency between A and C, or they may not. Whether or not an indirect dependency is created is a function of the directions given to the connections between A and B and between B and C. If both edges are directed into B as in the lower right-hand graph, then no indirect dependency is created. If the lower right-hand graph describes the causal dependencies, then information about A when added to information about B, should give further information about C, and information about C, when added to information about B, should give further information about A. In other words, A and C should be statistically dependent conditional on any set of variables that does not contain A and C but does contain B. Moreover, the converse seems true as well, and we therefore offer

> **Principle II:** if A and B are directly causally dependent and B and C are directly causally dependent, but A and C are not, then:

B is causally dependent on A, and B is causally dependent on C if and only if A and C are statistically dependent conditional on any set of variables containing B and not containing A or C.

The TETRAD II procedure uses the partial correlations supported by a sample to obtain an undirected graph using Principle I. We plan to implement a procedure that will apply Principle II to the undirected graph and the data to obtain the admissible orientations. The result will be a set of directed graphs; the size of the set will depend on the structure of the data, but it will always be enormously smaller than the space of *a priori* possibilities.

## 9.4 Examples

On large samples (n=2000) preliminary simulation tests of the procedures indicate that they are reliable. The procedure that applies Principle I has a tendency to underfit, that is it tends to omit undirected edges corresponding to causal dependencies in the structure from which the data were obtained. These omissions can reliably be recovered by submitting the models produced by the program to an elaboration procedure that adds edges whenever possible to eliminate the implication of independence constraints that are not satisfied in the data.

Data for a sample of 2000 population units were generated by Monte Carlo methods from a linear model with the causal structure shown in Figure 1 (normally distributed variates).

The data were then given to TETRAD II, which produced the undirected graph shown in Figure 2.

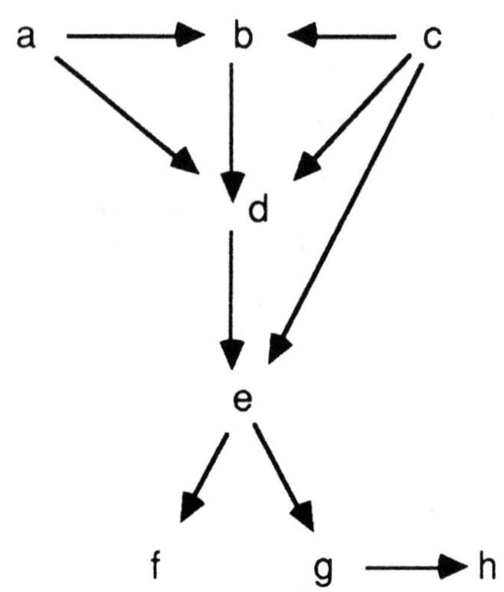


**Figure 1**

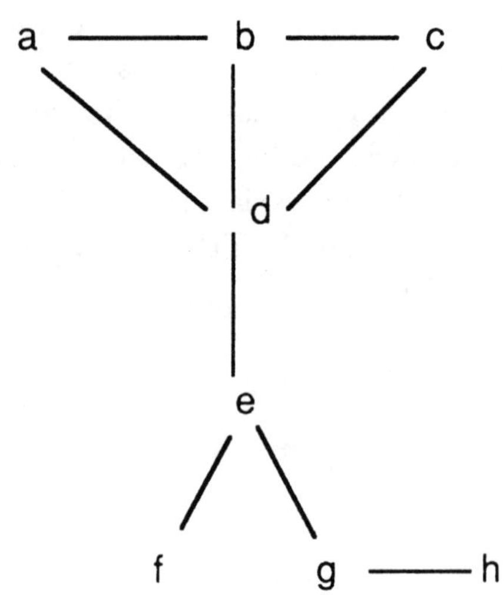


**Figure 2**

Principle II permits only two orientations of this graph, including the initial model used to generate the data. The omitted edge can be recovered by using elaboration techniques analogous to those of the TETRAD II program described in [Sprites 89a].

An empirical example is provided by recent work of Rogers and Maranto [Rogers 89]. They studied a number of theoretical accounts of the determinants of publishing productivity in psychology, and compared these accounts with original survey data they obtained. They first performed path analyses of six alternative models taken from the social science literature. They then formed a combined model that included all causal dependencies occurring in any of these models as well as two further dependencies. After estimating and testing the combined model, they eliminated the dependencies found not to be statistically significant. Their result is the causal model shown in Figure 3 (with coefficients estimated assuming linear dependencies)[5]

If we give the correlations from Rogers and Maranto's survey data to the procedure (whose present implementation assumes normal variates), we automatically obtain the undirected graph shown in Figure 4 *without imposing any prior substantive restrictions.*

The application of Principle I to the sample data yields eight of the eleven connections Rogers and Maranto postulate, and the three omitted edges are those which, in the linear model, have the smallest coefficients.

The time order of ability, graduate school quality, quality of first job, and publication in graduate school and other variables uniquely determines the directions that can be given to the edges in this undirected graph.

---

[5]GPQ is a scale formed from indicators of the quality of graduate programs; QFJ is a measure of the quality of the subject's first job; PREPRO is a measure of publications while in graduate study; PUBS a measure of publications since leaving graduate school, and CITES indicates frequency of citation the subjects scholarly works.

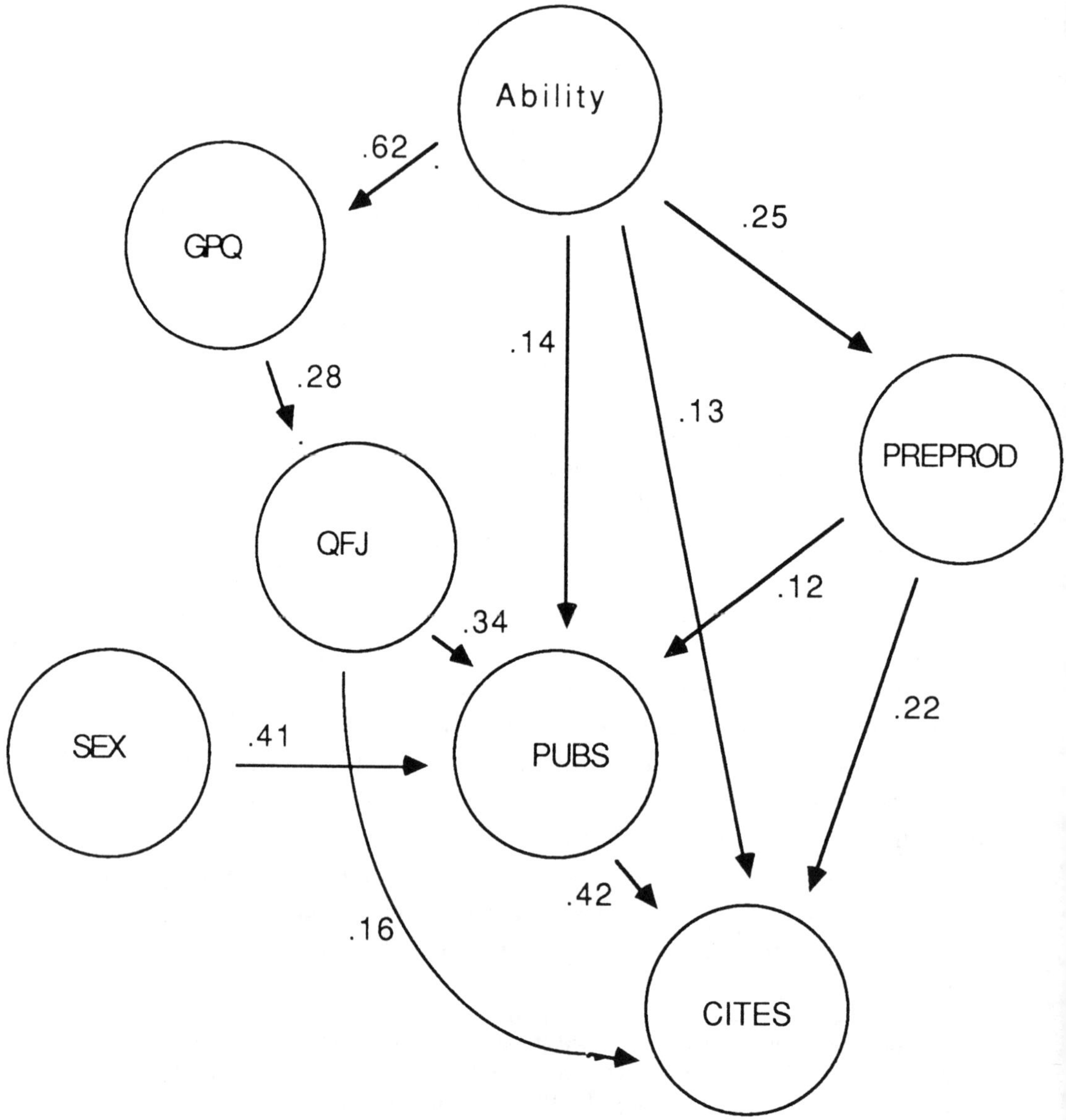
Ability
.62
.25
GPQ
.14
.13
.28
PREPROD
QFJ
.12
.34
.22
SEX
.41
PUBS
.42
.16
CITES

**Figure 3**

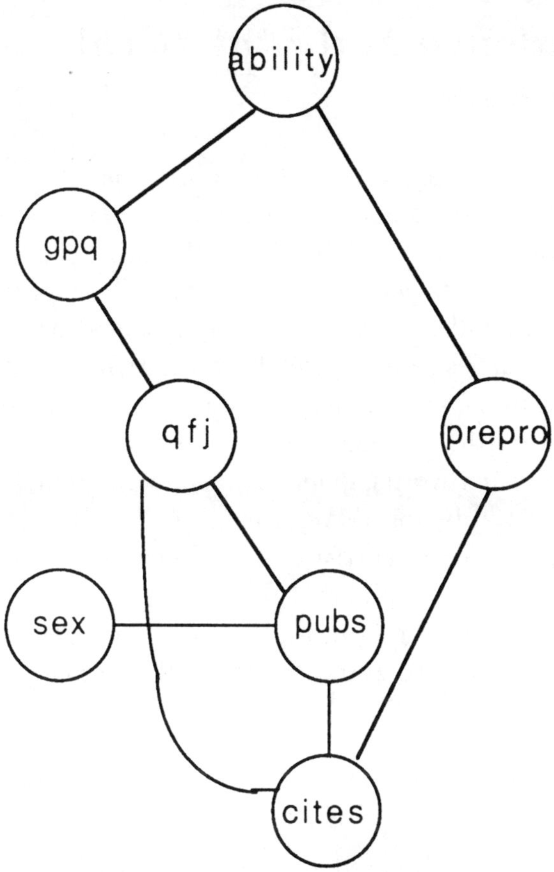


Output of Measured

**Figure 4**

In this case the sample size is not large and we used a test for vanishing partial correlations that assumes normal variates even though some of the variables in the Rogers and Maranto study certainly are not normally distributed. The result is that we obtain many more vanishing partial correlations than seem plausible, and therefore the application of Principle II failed to determine the orientation of the edges in this case. Thanks to the time order, however, the orientation is uniquely fixed.

## 9.5 Further Connections Between Causal Dependence and Statistical Dependence

Principles I and II determine from a probability distribution a set of alternative causal structures based on the relations of statistical dependence and independence in the distribution. The principles do not generally determine a unique causal structure. Conversely, given a causal structure, the two principles determine some of the statistical dependencies and independencies in any probability distribution represented by that causal structure, but generally a causal structure does not determine all of the statistical dependencies in probability distributions that it represents. In other words, given an acyclic directed graph, Principles I and II do not tell us for all vertices x, y and all sets of vertices S, whether or not x and Y are statistically independent conditional on S. There is a further principle, duc to J. Pearl and his colleagues [Pearl 1988], relating causal directed graphs and probability distributions that does give that information. To state it we need some definitions:

> An **undirected path** in an acyclic directed causal graph is a sequence of vertices $v_1, ..., v_n$ such that for all i between 1 and n-1, $v_i$ and $v_i + 1$ are directly causally connected.
>
> A vertex $v_j$ on an undirected path is a **collider on that path** if and only if the graph contains the directed edges $v_{j-1}- > v_j$ and $v_{j+1}- > v_j$.
>
> A **directed path** from $v_1$ to $v_n$ is an undirected path such that for all i between 1 and n-1, the graph contains the directed edges $_i- > v_{i+1}$.
>
> A vertex y is a **descendent** of a vertex x if and only if there is a directed path from x to y.
>
> Following Pearl, we say that variables x, y are **d-separated** by set S if and only if there exists no undirected path between x and y, such that (i) every collider on that path has a descendent in S and (ii) no other vertex on the path is in S.

With these definitions we can state Pearl's principle:

> **Principle III:** A directed acyclic graph represents a probability distribution on the variables that are vertices of the graph if and only if:
>
> for all vertices x, y, and all sets S of vertices in the graph, S d-separates x and y if and only if x and y are independent conditional on S.

Principle III is not immediately intuitive, and we do not have space to enter into issues concerning its justification, save to observe that it agrees with many intuitive principles about connections between causality and statistical dependence, and that it is satisfied by all of the usual modelling formalisms, e.g., in linear structural equation models. The connection between Principle III and Principles I and II may be unclear, however, and we report the following theorems.

> **Theorem 1:** Let P be a probability distribtuion represented by an acyclic directed graph G according to Principle III. Then G is an orientation of an undirected graph U that represents P according to Principle I.
>
> **Theorem 2:** Principle III implies Principle II.
>
> **Conjecture:** Let Γ be the set of directed graphs that represent probability distribution P according to Principle III. The Γ is also the set of directed graphs obtained from P by Principles I and II.

Principle III, together with these two theorems and two other results to be stated later that use the Principle, has some remarkable implications.

## 9.6 Characterizing Statistically Equivalent Models

There is a small literature on the notion of "statistically equivalent models." [Lee 1987]. A few writers have attempted to give sufficient conditions for two models to be statistically equivalent. The notion is not, however, very well characterized. Intuitively, the idea is that models are statistically equivalent with respect to a set of measured variables if no statistical inference based on sample data can reliably decide between them, no matter how large the sample. In practice, however, statistical equivalence has sometimes been characterized with respect to a particular estimation procedure or test, and more often not at all.

We will say that two directed graphs sharing a common set of vertices representing measured variables are **statistically equivalent** if the graphs contain the same number of directed edges and if, according to Principle III, the graphs imply the same collection of constraints on the measured variables.

It should be noted that for graphs in which there are no "latent" or unmeasured common causes, this condition implies that two graphs are statistically equivalent if and only if they represent exactly the same set of statistical dependencies and independencies. For graphs with latent variables this is not the case, since the constraints amongst the measured variables that such graphs imply (in virtue of the statistical dependencies and independencies they imply for all variables, both latent and measured) are not confined to statistical dependencies and independencies.

An immediate corollary to theorems I and II and to the conjecture is that Principles I, II and III provide a more or less feasible[6] procedure for determining all of the models statistically equivalent to any given model without latent variables. The procedure takes the given graph, forms its undirected graph, and calls Principle III whenever the application of Principle II requires a fact about statistical dependence. This procedure will be implemented in the TETRAD II program. The more general problem of finding a feasible procedure to characterize the class of models statistically equivalent to a given latent variable model has still to be solved.

[6]The procedure is clearly worst case exponential, and we guess that the problem of deciding statistical equivalence is NP hard. By "feasible" we mean a procedure that will work in reasonable time on available computers on models of the size one typically finds in social and behavioral sciences (exclusive of econometrics).

# 9.7 Inferring Latent Variables

## 9.7.1 Latent variables

Coupled with robust tests of vanishing partial correlations (or, for that matter, any robust test of statistical conditional independence) the procedures we have described provide a complete, almost distribution free, asymptotically reliable discovery procedure when all of the sources of statistical dependence between variables are due to causal relations amoung the measured variables. But what if they are not? What if there are unmeasured factors operating? Unmeasured variables are nothing mysterious. If one goes to any social science data bank one will find that variables measured in one data set are not measured in another. Unmeasured variables are just that – unmeasured; they need not be (and we expect generally are not) unmeasurable. It would seem that a central problem for scientific discovery in the social sciences is to find a reliable way to decide whether or not the statistical dependencies found amoung some of the variables in a sample are due to variations in variables that were not measured. We know of no reliable criterion for decisions of this kind in the social scientific or statistical literature. Work on Principle II yields one.

## 9.7.2 Tetrad constraints from acyclic directed graphs

The analysis of constraints implied by models goes back almost to the very origins of covariance analysis. Besides vanishing partial correlations of various orders, another kind of constraint, vanishing tetrad differences, has been frequently considered in the psychometric and social science literature. A tetrad constraint is just a vanishing difference of products of correlations involving four distinct variables:

$$\rho_{ij}\rho_{kl} = \rho_{ik}\rho_{jl}$$

When does a directed graph (with or without unmeasured variables) imply a tetrad constraint? In Glymour, Scheines, Spirtes and Kelly [1987] an algorithm was given to determine the tetrad constraints implied by a graph, assuming the variables are linearly related. But that is a very stringent assumption. Spirtes has recently found a necessary and sufficient condition for a probability distribution that is representable by a directed acyclic graph using Pearl's formalism (Principle III).

Using Pearl's condition, Principle III, Spirtes has recently given a necessary and sufficient condition for a graph to represent a distribution with a vanishing tetrad difference.

### 9.7.3 A sufficient condition for latent variables

A set of vanishing partial correlations can entail a vanishing tetrad difference. In linear causal models in which all of the variables are measured, all constraints implied by the graph of the model are axiomatized by the statistical dependencies and independencies represented by the graph, and all of the implied vanishing tetrad differences are implied by vanishing partial correlations. This observation led us [Glymour 1987] to suggest a heuristic for introducing latent variables: if a tetrad constraint holds that is not implied by vanishing first-order partial correlations amoung the measured variables, introduce latent variables. The theoretical grounds for a generalization of this heuristic can now be obtained rigorously as a corollary of Spirtes' characterization of the representation of vanishing tetrad differences:

> **Theorem III:** Suppose multinormal probability distribution P implies a vanishing tetrad difference
>
> $$\rho_{ij}\rho_{kl} = \rho_{ik}\rho_{jl}$$
>
> Then a directed acyclic graph G of a linear model represents P in accordance with Principle III only if there exists a set S of vertices of G such that the variables i and j are independent conditional on S, and so are the pairs of variables k, l and i, k and j, l.

When a vanishing tetrad constraint is judged to hold in the population, but there is no set of *measured* variables that the variable pairs i, j, and k, l, and i, k, and j, l are conditionally independent on, it does not follow that the distribution on the measured variables is generated from a probability distribution including latent variables. What follows is that if the distribution on the measured variables is a restriction of any distribution represented by an acyclic directed graph in accordance with Principle III, then that directed graph must have latent variables.

## 9.8 Asymptotic Reliability

The asymptotic properties of procedures that infer causal structure from constraints exemplified in sample data depend on two things.

- First, on the asymptotic properties of the procedures for determining that a set of constraints is satisfied;
- Second, on the relation between directed graphs and sets of constraints satisfied in a population.

Asymptotically distribution free tests are available both for vanishing tetrad differences and vanishing partial correlations. Sample correlations may, however, be correlated as may sample tetrad differences, so that the properties of decisions made on a series of individual tests from samples may be different from the properties of decisions made simultaneously for collections of constraints. We doubt that this is a serious problem in the limit, but we have no proof.

We have also proved a theorem that, informally expressed, states that for any "natural" probability distribution on the linear coefficients and variances of linear models based on a given causal graph, the likelihood is zero that there will be a tetrad or other independence constraint that holds because of the coefficients and variances rather than because of the causal structure.

## 9.9 Conclusion

We believe the results described here, together with the TETRAD II results reported elsewhere, take a long step towards automated causal inference. However, a lot of work remains before a feasible robust general tool is available. The algorithm using Principle II needs to be implemented; the criterion for judging the presence of latent variables must be implemented; an automatic elaboration procedure using the strategy of TETRAD II needs to be implemented; more robust statistical tests of conditional independence need to be introduced; and the entire package needs to be tested extensively on simulated data. Combined with TETRAD II procedures for finding elaborations of an initial incomplete latent variable model, and with commercial packages for estimation and testing under a variety of distributional assumptions, major parts of the process of model construction can be automated, and automated with guarantees of reliability.

There is an important gap in the work we have described. Although we possess procedures for specifying causal models when statistical dependencies are not due to unmeasured common causes, and we possess means to determine when common causes are acting, and we possess means of finding omitted causal dependencies in a partially specified latent variable model, we do not possess a reliable informative procedure for constructing initial latent variable models from the data alone, nor do we possess a solution to the closely connected problem of feasibly generating all latent variable causal models statistically equivalent to a given model. These problems will require further fundamental research.

## References

Duncan, O. D., [1975]: *Introduction to Structural Equation Models*, Academic Press. New York.

Glymour, C., Scheines, R., Spirtes, P., and Kelly, K., [1987]: *Discovering Causal Structure: Artificial Intelligence, Philosophy of Science, and Statistical Modelling*, Academic Press. San Diego.

Hilary, F., and Palmer, E., [1973]: *Graphical Enumeration*, Academic Press. New York.

James, L.R., Mulaik, S.A., and Brett, J., [1982]: *Causal Analysis*, Sage Publications.

Joreskog, K., and Sorbom, D., [1984]: *LISREL VI: Users's Guide*, Scientific Software. Mooresville, IN.

Lee, S., [1987]: Ph.D. Thesis, Ohio State University.

Loehlin, J., [1987]: *Latent Variable Models*, Lawrence Ehrlbaum

Associates, Hilldale, NJ.

Pearl, J., [1988]: *Probabilistic Reasoning in Intelligent Systems: Networks of Plausible Inference*, Morgan Kaufmann Publishers Inc. San Mateo, Ca.

Rodgers, R.C., and Maranto, C.L., [1989]: "Causal Models of Publishing Productivity in Psychology", *Journal of Applied Psychology, Vol 74, No. 4*, 636-649.

Skyrms, B., [1980]: *Causal Necessity*, Yale University Press. New Haven.

Spirtes, P., Scheines, R., and Glymour, C., [1989 forthcoming]: "Simulation Studies of the Reliability of Computer Aided Model Specification Using the TETRAD II, EQS, and LISREL Programs", in *Sociological Methods and Research*, Sage.

Spirtes, P., [1989]: "A Necessary and Sufficient Condition for Conditional Independencies to Imply a Vanishing Tetrad Difference", *Technical Report Number CMU-LCL-89-3*, Department of Philosophy, Carnegie Mellon University, Pittsburgh, PA, 15213.

Suppes, P., [1970]: *A Probabilistic Theory of Causality*, North-Holland, Amsterdam.

# Chapter 10

# Modelling Conceptual Change

Paul Thagard,
*Cognitive Science Laboratory, Princeton University, 221 Nassau St., Princeton, NJ 08542. Email address: pault@clarity.princeton.edu*

## 10.1 The Problem of Revolutionary Conceptual Change

Scientific knowledge often grows slowly with gradual additions of new laws and concepts. But sometimes science undergoes dramatic conceptual changes when whole systems of concepts and laws are replaced by new ones. By analogy with political upheavals, such changes are called *scientific revolutions* [Cohen 1985, Kuhn 1970]. Although many historians and philosophers of science have stressed the importance of scientific revolutions, there has been little detailed analysis of the nature of such changes. How do conceptual revolutions occur? How can a new conceptual scheme arise and supplant an old one? What are these conceptual schemes whose transformation is so fundamental to scientific development?

I have been attempting to answer these questions from a viewpoint that is psychological and computational. This attempt is an application of an enterprise that I have called *computational philosophy of science*, which applies techniques derived from the field of artificial intelligence (AI) to describe the structure and growth of scientific knowledge [Thagard 1988]. My earlier book sketched computational accounts of problem solving, explanation, and hypothesis formation and evaluation. My more recent research extends the kind of account given there to more dramatic phenomena in the history of science: scientific revolutions. This chapter is an overview of the results so far of the project of developing a computational theory of conceptual change. I shall first quickly discuss the significance of an understanding of conceptual change for philosophy and artificial intelligence. Then, before describing my own approach to the problem, I shall give a brief critique of the very different quasi-computational approach to understanding science offered within the framework of formal learning theory. The description of my own approach summarizes the much more detailed analyses that have been offered elsewhere [Thagard, in press-a; Thagard in press-b; Thagard and Nowak, in press; Thagard 1989].

## 10.2 Philosophical Significance of Conceptual Change

I begin by quickly reviewing the importance of an understanding of scientific revolutions to the philosophy and history of science. For several decades prior to the 1962 publication of Kuhn's *Structure of Scientific Revolutions*, philosophy of science had been dominated by the logical empiricist approach. Exemplified by such philosophers as [Carnap 1936] and [Hempel 1965], the logical empiricists used the techniques of modern formal logic to investigate how scientific knowledge could be tied to sense experience. Like the contemporaneous views of [Popper 1959], the logical empiricists were largely ahistorical and apsychological, emphasizing the logical structure of science rather than its development.

Kuhn, along with other historically inclined philosophers such as [Hanson 1958] and [Feyerabend 1965], charged the logical empiricists with historical irrelevance. [Kuhn, 1970, p. 1] wrote: "History, if viewed as a repository for more than anecdote or chronology, could produce a decisive transformation in the image of science by which we are now possessed." There are many reasons to question both Kuhn's general view of scientific revolutions and his accounts of particular scientific episodes, but his general claim that the development of scientific knowledge includes revolutionary changes can be substantiated.

Kuhn's claims about scientific revolutions have caused great consternation among philosophers of science because of their apparent implications of irrationality. Kuhn so stressed the dramatic and non-continuous nature of scientific change that transitions in scientific theories or "paradigms" took on the appearance of cataclysmic, non-rational events. According to Kuhn's early statements, later moderated, a scientific revolution involves a complete change in standards and methods, so that rational evaluation of competing views using external standards appears impossible. He even said that when one theory or "paradigm" replaces another, the world changes. But a finer-grained theory of revolutionary change than Kuhn presented need not succumb to irrationalism. To get such a theory, however, we need analytical tools different from both the formal ones of the logical empiricists and the vague historical ones of Kuhn. Artificial intelligence offers the possibility of developing such tools.

## 10.3 Computational Significance

In cognitive science, computational models are the central technique for theoretical analysis. For many researchers, it is not just that thinking can be modeled computationally like the weather and other phenomena: thinking *is* a form of computation. To model mental structures and processes, programs are designed with data structures corresponding to the postulated mental structures and with procedures corresponding to the postulated mental processes. Running the program and doing psychological experiments provide ways of judging whether the model corresponds to psychological reality.

Artificial intelligence offers numerous tools for constructing these kinds of cognitive models. The AI sub-field of *knowledge representation* is concerned with techniques of representing information in a computer for intelligent processing. AI work on *problem solving* and planning is highly relevant to the thinking scientists that involves explanation and problem solving. *Machine learning* is the AI sub-field concerned with how computational systems can improve their performance by acquiring and modifying their structures and procedures.

We cannot, however, simply take over existing ideas from these fields. Although much work in machine learning has been done on topics such as concept formation, we shall see that available techniques are not adequate to account for the origins of many important scientific concepts. Moreover, AI researchers have concentrated on cases of learning by accretion of knowledge, rather than on cases of revolutionary replacement of complexes of concepts. New computational ideas are required to account for this kind of replacement. Previous work on scientific discovery has neglected conceptual change (for example, [Langley, Simon, Bradshaw, and Zytkow 1987]).

## 10.4 Formal Learning Theory

There is more than one possible route to the integration of philosophy of science with computational ideas. I have been espousing a method that integrates historical philosophy of science with mainstream AI research on knowledge representation and learning. In contrast, a different approach as been proposed that integrates philosophy of science more in the style of the logical positivists with *formal learning theory*, an enterprise that has developed within the context of theoretical computer science [Osherson and Weinstein 1989; Kelly and Glymour 1989]. In order to assess the relevance of this kind of approach to understanding the development of scientific knowledge, I shall translate the somewhat daunting formalism of formal learning theory into a description of a program that purportedly models a scientist. We shall then be able to evaluate the verisimilitude of formal learning theory as a model of science.

The four authors just cited discuss science within the tradition of [Gold 1967] in which the goal is to *identify* a language given a string of inputs. Analogously, a scientist is conceived as a function that, given a sequence of formulas representing observations of the environment, produces a set of recursively axiomatizable formulas that represent the actual structure of the world. All formulas belong to a first-order language L. Thinking of the formal scientist now as a computer program CP, rather than as an abstract function, we can reinterpret the formalism as follows. The *input* to CP consists of

(a) Sequences of basic formulas of L, i.e. atomic or negated atomic formulas, and

(b) A structure R in the sense of formal semantics, i.e. a domain D and a set of relations (sets of n-tuples) of the elements of D.

The *output* of CP consists of recursively axiomatizable subsets of L. The goal of CP is eventually to reach a point such that it outputs a formula of L if and only if the formula is true in R, irrespective of which sequences of basic formulas are input. For the full formal specification of these notions, the reader should consult [Kelly and Glymour 1989] and [Osherson and Weinstein 1989], who go on to prove theorems concerning the conditions under which the goal of CP is effectively realizable. My informal characterization should suffice, however, to evaluate the extent to which this formal model of a scientist is unrealistic about actual science. Of course, any idealization will deviate from reality, and no one can doubt the efficacy of scientific idealizations such as those found in Newtonian mechanics. But we must also be wary of cases, such as much of economic theory, where idealizations are so far from reality that their value lies more as mathematical exercises than as descriptions or explanations of the phenomena.

I see four major problems with the picture of science that this formal approach embodies. First, from the perspective of conceptual change, the fixity of the language L is completely problematic. One of the most interesting aspects of the development of science is the introduction of new terms like *oxygen*, *electron*, *quark*, and *gene*. Terminological innovation is not confined merely to theoretical science, since experimental developments also bring with them new terms such as *telescope* and *microscope*. From an AI point of view, a central aspect of the problem of conceptual change is to specify procedures that can enhance the language L by constructing new concepts. In addition, except perhaps for physics, the restriction of L to a first-order language is unrealistic. In the social sciences, explanations of other people's behavior often

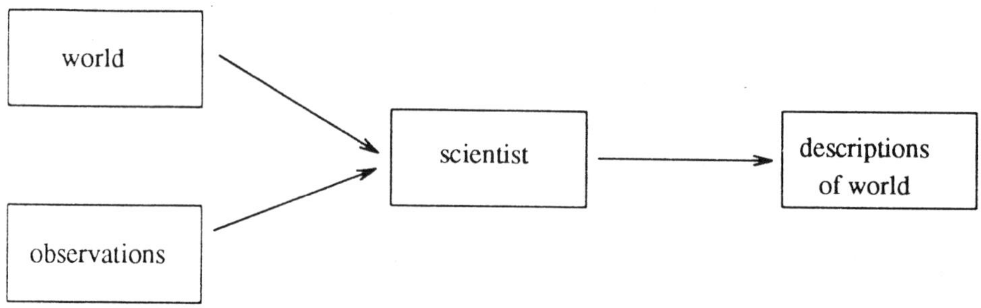


**Figure 1:** The model of the scientist in formal learning theory.

involves generalizations about their beliefs, desires, and other propositional attitudes. A language rich enough to represent these generalizations will have be able to quantify over propositions, not just objects as in first order languages. Biology also appears to need a higher-order logic, since discussions of evolution often quantify over qualities, features, and traits of organisms.

The second major problem with the formal learning theory model of science is that the output of the program CP - what the scientist learns - is a function only of the input from the environment. But in science, the generation of new concepts and theories is based on accumulated knowledge as well as on new input from the environment. Formal learning theory assumes a processing system like that depicted in Figure 1, while Figure 2 presents a more realistic portrayal. The laws, theories, and concepts generated by scientists play a role in generating further laws, theories, and concepts; experiments that lead to new observations also play a role, but only as mediated by the scientist's existing belief system. From a positivist perspective, it may make sense to model a scientist as a function from sequences of basic formulas representing the environment to sets of formulas representing scientific knowledge, but this model bears little relation to what occurs in science. A rich model will have to take seriously the problems of how scientists' knowledge is represented and how it can be expanded. The formal learning theory approach assumes that all scientific knowledge is represented by sets of formulas, but we shall see that conceptual organization is also

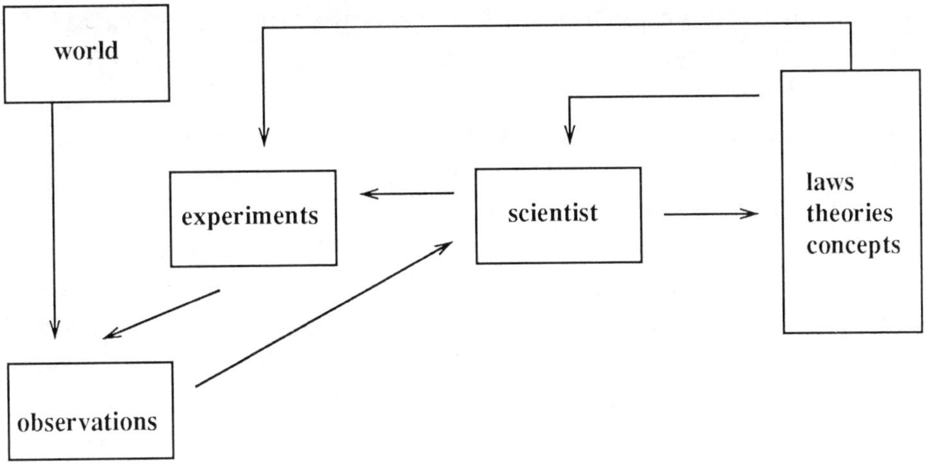


**Figure 2:** A more realistic model of a scientist.

important.

Gold's identification notions used by Kelly, Glymour, Osherson, and Weinstein are also unrealistic in expecting a scientist not to make mistakes. In real life, of course, the data are sometimes erroneous, and scientists should not be expected to get everything right. The learning model of [Valiant 1984] is superior in being probabilistic about success, but shares with identification models the unrealistic abstraction from the effects of existing knowledge in the formation of new knowledge.

The third major problem with formal learning theory is that it mistakenly takes the central aim of science to be the accumulation of true sentences. The greatest successes of science are not constituted in terms of achieving truths and avoiding falsehoods, but rely on the accomplishment of great *explanatory* theories like Lavoisier's oxygen theory, Darwin's theory of natural selection, and Einstein's theory of relativity. Some truths are much more important than others. I will refer below to my computational theory of explanatory coherence that shows how theories can be evaluated. The goal of CP is to generate formulas true of the world, but in theoretical science we never have any way of determining whether our theories are true independent of their explanatory power. Formal learning theory's criterion of success assumes that a function's ability to generate true sentences can be assessed, but in science such an assessment is never possible. We can have very good grounds for accepting a theory on the basis that it provides a better explanation of the evidence than its competitors, but that is the closest we can come to an assurance that the theory is true.

The fourth major problem with the identification approach to formal learning theory is that its results concern *effective* procedures which may not necessarily be computationally tractable. In contrast, Valiant's learning theory [Valiant 1984] proves results about what can be learned in polynomial time. Even that restriction is too loose, since learning results that involve large polynomials such as $n^{100}$ are unrealistic as models of scientists if $n$ gets at all large. To be interesting, theoretical results about the development of scientific knowledge should concern systems - such as human scientists - that are capable of operation in real time. I estimate that more is likely to be learned about such systems by the attempt to build them than by the analysis of unrealistic formal models.

I am certainly not opposed to formal analyses in principle. We have been fortunate in this century to have seen formal development of genuine epistemological interest, such as Gödel's theorem about the incompleteness of logical systems for mathematics, Turing and Church's work on effective computability, and Cook's initiation of the theory of NP-completeness. All three of these results combined mathematical virtuousity with relevance to important questions about computation. So far, the research applying formal learning theory to the philosophy of science has been impressive as an exercise in mathematical logic but unilluminating about science.

## 10.5 Theses on Conceptual Revolutions

My own approach to understanding conceptual change can be summarized by six theses that constitute tentative claims about what is required for a computational theory of conceptual revolutions. First, we need to understand the kinds of structures undergoing transformation. Thesis 1 accordingly states:

1. *Scientific revolutions all involve major transformations in conceptual and propositional networks.*

To specify the transformations it is necessary to describe the structure of conceptual and propositional networks. How are concepts organized? It is common in artificial intelligence systems to have much of the organization provided by KIND or IS-A hierarchies. For example, Tweety is a canary, which is a kind of bird, which is a kind of animal, which is a kind of thing. Similarly, psycholinguists have noticed the importance of kind hierarchies for organizing the mental lexicon, and in addition have emphasized the role that *part-whole* hierarchies play. I therefore conjecture:

2. *Conceptual networks are primarily structured via kind and part-whole hierarchies.*

From theses 1 and 2 follows the conjecture that all scientific revolutions involve transformations in kind and/or part-whole relations. A thorough analysis of the chemical revolution brought to light major changes in both kind and part-whole relations [Thagard in press-b]. Similar analyses that have been performed for the Copernican, Newtonian, Einsteinian, quantum, Darwinian, and geological revolutions support thesis 2.

Empirical laws are usually framed in the same vocabulary as the observational descriptions on which they are based. In contrast, theories often invoke entities and processes that are unobservable or at least unobserved. I have elsewhere contended that theoretical concepts can be formed by *conceptual combination*, in which new concepts, as represented by frames in the sense of [Minsky 1975], can be formed from old ones by using heuristics to select slots from existing concepts [Thagard 1988; Holland, Holyoak, Nisbett, and Thagard, 1986]. I therefore conjecture:

3. *New theoretical concepts generally arise by mechanisms of conceptual combination.*

Theses 2 and 3 deal with the structure and origin of conceptual networks, while 4 and 5 address the same questions for propositional networks. I have developed a computational theory of explanatory coherence designed to account for revolutionary and non-revolutionary cases of theory evaluation [Thagard 1989]. On this account, the most important relations between propositions concern explanation and contradiction. On the basis of that theory, I conjecture:

4. *Propositional networks are primarily structured via relations of explanatory coherence.*

Where do new hypotheses come from? In the nineteenth century, the philosopher C. S. Peirce coined the term *abduction* to refer to the formation of explanatory hypotheses, and computational research on abduction is growing rapidly (for example, [Thagard 1988]). I conjecture:

5. *New theoretical hypotheses generally arise by abduction.*

Revolutions involve such major transformations in conceptual and propositional organization that there is no way in which simple techniques for operating on initial representations could have brought about the change gradually. Rather, a new theoretical and conceptual framework must be developed somewhat independently of the existing one. I conjecture:

6. *The transition to new conceptual and propositional networks occurs because of the greater explanatory coherence of the new propositions that use the new concepts.*

My theory of explanatory coherence has been implemented computationally and applied to show how Lavoisier's argument for the superiority of his oxygen theory over the accepted phlogiston theory, and Darwin's argument for evolution by natural selection, can be understood in explanatory coherence terms. A similar analysis has been applied to the geological, Copernican, and Newtonian revolutions, and to the debate about why dinosaurs became extinct [Thagard 1989; Thagard and Nowak 1988, in press-a; Nowak and Thagard forthcoming; Thagard forthcoming-b].

Although the historical support for these conjectures has been rapidly accumulating, they should still be considered as hypothetical. I have most doubts about theses 3 and 5, which claim that the mechanisms of conceptual combination and abduction suffice for concept and hypothesis formation. Although both of these mechanisms have been implemented computationally in the system PI [Thagard 1988], I have so far simulated only simple cases of concept and hypothesis formation. I would not be surprised if more complex mechanisms are needed. Contrary to [Kelly and Glymour 1989, p. 220], I see no point in making "the logic of discovery look, at last, like logic". Understanding of the development of scientific concepts and hypotheses requires attention to complex mechanisms for producing scientific concepts and hypotheses that go far beyond the resources of formal logic.

## 10.6 Kinds of Conceptual Change

A full defense of theses 1-6 will require a book. In the rest of this chapter, I shall merely outline the historical support for the implication of theses 1 and 2 that scientific revolutions involve transformations in the kind-hierarchies and part-hierarchies that organize conceptual systems. I shall provide a taxonomy of kinds of epistemic change, distinguishing conceptual change from belief revision, and give examples of the kinds of conceptual change that occur in scientific revolutions.

Figure 3 is a taxonomy of kinds of epistemic change. The sort of change most often discussed by philosophers has been belief revision, involving the addition or deletion of beliefs. Elsewhere I have argued against the contention that conceptual change reduces to belief revision [Thagard forthcoming-a]. The thrust of the argument is that although we can translate kinds of conceptual change into belief revision, lost in the translation are important facts about the structure of knowledge and the processes that produce it. Belief revision and conceptual change go hand in hand, but a system must have a mechanism for producing a new concept (predicate) before it can use that concept in new beliefs. For example, to simulate the development of Lavoisier's theory, one would need to have a mechanism for forming the predicate *oxygen* in order to be able to represent beliefs such as that things burn by combination with oxygen.

Addition of concepts is particularly important when it modifies the

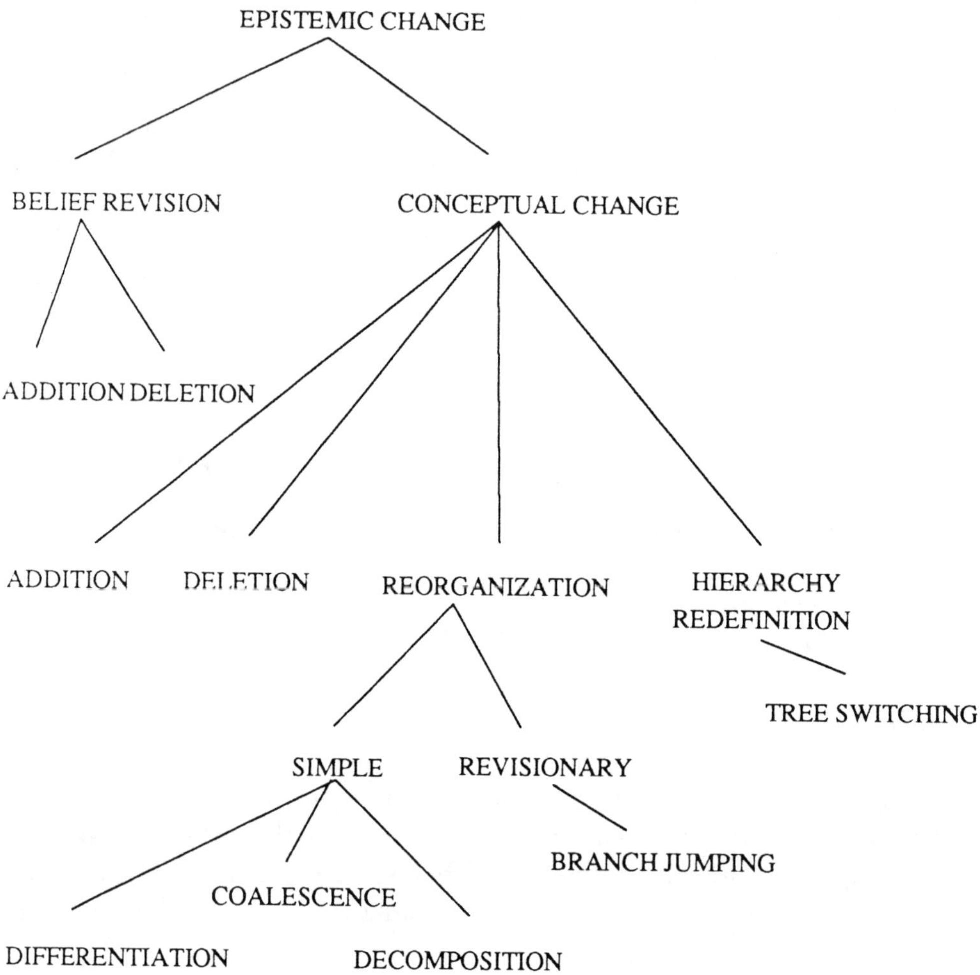


**Figure 3:** Taxonomy of conceptual changes.

hierarchies that organize the conceptual system. [Carey 1985] discusses two such kinds of conceptual change: coalescence, when one adds a new superordinate kind that links two things previously taken to be distinct; and differentiation, when a distinction is made producing two kinds of things. She illustrates coalescence by the conceptual change that occurs in children when they form the concept of *alive* that has as subordinates both animals and plants. Her prime example of differentiation concerns heat and temperature, which were taken to be the same until Black distinguished them. Conceptual change occurs in the part-hierarchy when parts are discovered, as has often in occurred in physics with the discovery of molecules, atoms, and sub-atomic particles.

Figure 3 characterizes coalescence, differentiation, and decomposition as *simple* conceptual reorganizations, since they involve local alteration in kind-relations or part-relations, and the alterations are additions without deletions. In scientific revolutions, however, it often happens that whole systems of kind-relations or part-relations are replaced. For example, in the chemical revolution, the almost-modern taxonomy of kinds of substances produced by Lavoisier was strikingly different from the taxonomy that was central to the phlogiston theory, which had three kinds of earth including phlogiston as fundamental principles, and substances such as gold and lead as mixtures. This revolution also involved a dramatic change in part-relations, since in the phlogiston theory metals were thought of as complexes consisting of phlogiston plus calxes, where a calx is a simple substance corresponding to what we now call an oxide. I call these kinds of conceptual changes *branch jumping*, since they involve the movement of a concept from one branch of the kind-hierarchy or part-hierarchy to another.

Figures 4 and 5 show a striking case of branch jumping. Figure 4 should strike the reader as a rather odd classification of celestial bodies, in contrast to the modern view roughly shown in Figure 5; but the figure represents the views of Ptolemy's *Almagest* that dominated astronomical thinking for over a thousand years. Copernicus brought about the reconceptualization of the earth as a kind of planet. For Ptolemy the sun and moon were planets, since they were thought to rotate round the earth just like Venus, Mars, and so on. But Copernicus conceived the moon as uniquely rotating around the earth, and Galileo's

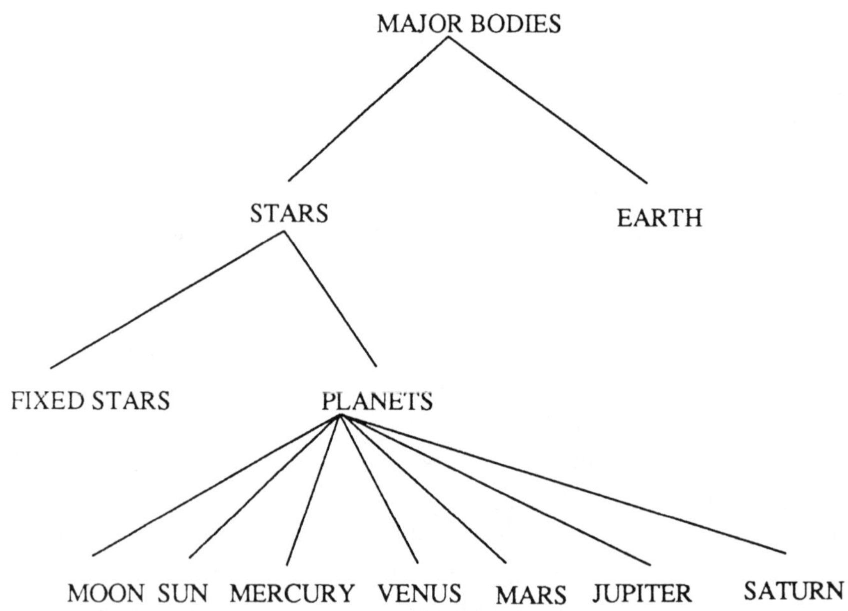


**Figure 4:** Partial conceptual system of Ptolemaic astronomy. The lines indicate kind relations.

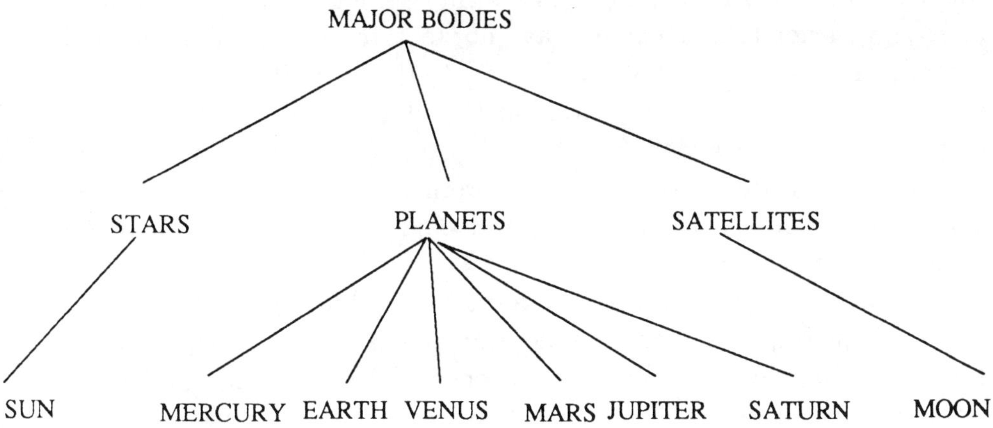


**Figure 5:** Partial conceptual system of current astronomy.

discovery of the moons of Jupiter helped to produce the generalized notion of satellite. The sun was not recognized as a star until around 1800. The move from the Ptolemaic to the Copernican system required the substitution of a whole system of beliefs that corresponded to a radically different system of conceptual organization.

As important as branch jumping is, it is not the most radical kind of conceptual change. By *tree switching* I mean cases where the very principle of organization of concepts changes: the hierarchical tree gets fundamentally redefined. I have not identified any such changes in the Copernican, Newtonian, or chemical revolutions, but they appear to have occurred in four other major revolutions. Darwin initiated a major alteration in the concept of *kind*, since species became understood in historical terms rather than in terms of similarity of features. This change introduced a new principle of classification: two species are of the same kind if they are descended from a common ancestor, irrespective of their differences in surface appearance. Relativity theory and quantum theory introduced changes in our understanding of part-hierarchies. The relativistic concept of a space-time continuum undercuts everyday notions of spatial and temporal parts and requires rethinking them in terms of an integrated notion of space-time. In quantum theory, Heisenberg's uncertainty principle can be interpreted as saying either that a particle has no definite position until it is observed, or less idealistically as saying that its position is inherently statistical. Either way we abandon the classical common-sense view of particles having a definite place in space and time. Similarly, the notions of part and whole become more indeterminate than they were classically, since spatial and temporal parts are no longer absolutely determinable.

I hope that these examples suggest to you that conceptual organization is an important epistemological question not to be ignored in favor of belief revision. Without a sense of how concepts and beliefs are organized, it becomes impossible to see why the transition from one theory to another is as difficult as the history of science shows it to be. Transitions, are not, however, so difficult as Kuhn suggested: scientific revolutions are *not* a matter of proponents of old beliefs dying off to be replaced by holders of totally new world views. In the major scientific revolutions since Newton, most scientists have adopted the new theory, although it often takes them years to make the transition. I offer the following picture of how the transition takes place:

1. A scientist with a theory embedded in a conceptual scheme becomes aware of a new theory that competes with the one already held.

2. Although initially skeptical, the scientist sets out to learn more about the new theory, and gradually accumulates its conceptual system and an understanding of its explanatory claims.

3. The scientists comes to appreciate that the new theory has greater explanatory coherence than the old one.

4. The old theory, and its attendant conceptual system, drop into disuse.

In this way, a theory of explanatory coherence can be the driving force behind theory replacement and major conceptual change. The crucial point to notice is that the new conceptual scheme does not arise by piecemeal modification of the old one. Rather, the new one must be built up largely on its own, and its replacement is the result of a global judgment of explanatory coherence.

## 10.7 Conclusion

My description of conceptual change in this chapter has been very sketchy, but the historical and computational ideas are worked out elsewhere. I hope it suffices, however, to give the flavor of the kinds of analyses of conceptual change that are possible within a framework that embraces historical philosophy of science and artificial intelligence. While lacking the rigor of analyses possible within formal logic and learning theory, the historical/computational approach has much greater promise for illuminating the development of scientific knowledge.

### References

Carnap, R., [1936/1937]: "Testability and meaning". *Philosophy of Science, 3*: 419-471; 4: 1-40.

Carey, S., [1985]: *Conceptual change in childhood.* MIT Press. Cambridge, Ma.

Cohen, I.B., [1985]: *Revolution in science.* Harvard University Press. Cambridge, Ma.

Feyerabend, P., [1965]: "Problems of empiricism". In Colodny, R., (ed.). *Beyond the edge of certainty.* 145-260, University of Pittsburgh Press. Pittsburgh.

Gold, E., [1988]: "Language identification in the limit". *Information and Control, 10,* 447-474.

Hanson, N.R., [1958]: *Patterns of discovery.* Cambridge University Press. Cambridge.

Hempel, C., [1965]: *Aspects of scientific explanation.* The Free Press. New York.

Holland, J., Holyoak, K., Nisbett, R., and Thagard, P., [1986]: *Induction: Processes of inference, learning, and discovery.* MIT Press/ Bradford Books. Cambridge, Ma.

Kelly, K., and Glymour, C., [1989]: "Convergence to the truth and nothing but the truth". *Philosophy of Science, 56,* 185-220.

Kuhn, T. [1970]. *Structure of scientific revolutions*, 2nd edn. University of Chicago Press. Chicago. First published 1962.

Langley, P., Simon, H., Bradshaw, G., and Zytkow, J., [1987]: *Scientific discovery.* MIT Press/Bradford Books. Cambridge, Ma.

Minsky, M., [1975]: "A framework for representing knowledge". In Winston, P.H., (ed.), *The psychology of computer vision*, 211-277. McGraw-Hill. New York.

Nowak, G., and Thagard, P., [forthcoming]: "Copernicus, Ptolemy, and explanatory coherence". For a volume of *Minnesota Studies in the Philosophy of Science.*

Osherson, D., and Weinstein, S., [1989]: "Identifiable collections of countable structures". *Philosophy of Science, 56* 94-105.

Popper, K., [1959]: *The logic of scientific discovery.* London: Hutchinson.

Thagard, P., [1988]: *Computational philosophy of science.* Cambridge, Mass.: MIT Press/Bradford Books.

Thagard, P., [1989]: "Explanatory coherence". *Behavioral and Brain Sciences, 12,* 435-467.

Thagard, P., [in press-a]. "Concepts and conceptual change". *Synthese.* February, 1990.

Thagard, P., [in press-b]. "The conceptual structure of the chemical revolution". *Philosophy of Science.* June, 1990.

Thagard, P., [forthcoming]: "The dinosaur debate: Explanatory coherence and the problem of competing hypotheses". To appear in a volume on philosophy and artificial intelligence edited by R. Cummins and J. Pollock.

Thagard, P., and Nowak, G., [1988]: "The explanatory coherence of continental drift". In Fine, A., and Leplin, J., (Eds.), *PSA 1988*, vol. 1. East Lansing, Mich.: Philosophy of Science Association, 118-126.

Thagard, P., and Nowak, G., [in press]: "The conceptual structure of the geological revolution". In Shrager, J., and Langley, P., eds., *Computational Models of Discovery and Theory Formation.* Hillsdale, NJ: Erlbaum.

Valiant, L., [1984]: "A theory of the learnable". *Communications of the ACM*, 27, 1134-1142.

# Chapter 11

# Plan Recognition in Stories and in Life*

Eugene Charniak and Robert Goldman
*Department of Computer Science, Brown University, Box 1910, Providence RI 02912*

Plan recognition does not work the same way in stories and in "real life" (people tend to jump to conclusions more in stories). We present a theory of this, for the particular case of how objects in stories (or in life) influence plan recognition decisions. We provide a Bayesian network formalization of a simple first-order theory of plans, and show how a particular network parameter seems to govern the difference between "life-like" and "story-like" response. We then show why this parameter would be influenced (in the desired way) by a model of speaker (or author) topic selection which assumes that facts in stories are typically "relevant".

*This work has been supported in part by the National Science Foundation under grants IST 8416034 and IST 8515005 and Office of Naval Research under grant N00014-79-C-0529.

## 11.1 Introduction

Plan recognition in stories does not work in the way it does in "real life." Consider:

> Jack wanted to kill himself. A pistol dating back to the days of Teddy Roosevelt and the Rough Riders was a family heirloom.

If you encountered this in a story you would seriously entertain the idea that Jack might try to shoot himself. If asked to quantify your belief you would say that the chances were less that one, but more than, say, .2. You would presumably say the same here:

> Jack wanted to kill himself. There was a rope in the closet.

Now, of course, Jack would be hanging himself.

Life works differently. If the facts in the first story were true of a friend of yours you would not really believe that he would use the gun. After all, he probably has rope, knives, pills, and a high building to jump from. (This is not to say that shooting might not occur to you, but simply that you would not believe that he would do it with anything like a 20% probability.) We assume that stories give us selected, typically relevant, information, while life gives us unselected, mostly tangential, facts. Somehow this affects plan recognition in the two areas.

This is not news to those of us working in plan recognition in stories. However, as it complicates things, and makes what we are doing seem more ephemeral, we steadfastly ignore the fact. Indeed, the only published reference to the problem I have found is [Kaplan, 1978]. Kaplan proposes a class of "language driven inferences" which are distinct from "domain driven inferences." However, since he does not specify what these are, this is really a restatement of the problem. We all agree that we reach extra conclusions in stories, and thus must be making extra inferences. The question is *why*? (Not to mention *which ones*, and *how*?)

We take the position that plan recognition is done the same way (in people) for both stories and life. Therefore we are looking for a single theory which can handle both, and explains why and how they differ. This chapter will present such a theory, at least for the case of how objects in a story influence our perceptions of people's plans in a way different from the object's presence in the world. In Section 2 we formulate the connection between plans and their objects in a simple first-order version of frames and slots. Section 3 then recast this in probabilistic terms, using Bayes networks. We will see how by setting a particular network parameter (the prior probability that two things are the same) to different values we can get both story-like and life-like interpretations out of the network, and thus in a crude way shows how a single system could handle both cases. Section 4 improves this theory by showing how reasonable assumptions about speaker's intentions, "only mention relevant things," will also modify this parameter. We thereby get the theory we want: one which handles both stories and "life", and explains the difference in terms of a simple assumption about selection in the case of stories.

## 11.2 Preliminaries

### 11.2.1 Objects and plans in a first-order frame theory

We adopt a simple model of plan recognition in which plans are complicated actions (ones with sub-steps) and all plans are built out of a complete library of plan schemas. Plan schemas are natural kinds ("frames") and the steps of a plan are "slots" in frames. In the first order reconstruction, these slots are functions from instances of a plan schema to the actions which accomplish the step in the plan. To take an example we will be using a lot, here is how we would say that some person j1 is engaged in a hanging plan h2, and that a particular getting event g3 filled the step in the plan by which the agent gets the necessary rope (which we will call the **get-step**):

```
(person j1)
(hang h2)
(= (agent h2) j1)
(get g3)
(= (get-step h2) g3)
```

The last line here serves as the explanation of the getting event in terms of a higher plan. Objects in plans are also connected to the plan via slots (functions). So the rope-of slot would associate a hanging plan with the rope used in it, as in (= (rope-of h2) r4). We assume that all slots have restrictions on what can fill them; rope-of must be filled by a rope. This is expressed as the general constraint (hang ?x)→(rope (rope-of ?x)) Throughout the rest of this paper we will be considering the plan recognition problems inherent in the two stories:

> "There was a rope in the closet. Jack killed himself."
> "Jack got a rope. He killed himself."

We take it for granted that the second example differs from the first in that in real life we would assume that Jack hung himself.

It is important to note that nothing we say is particular to these examples. So, more generally, note that if an object $j$ fills the slot $s$ in the instance $i$ of the plan schema $p$, where $s$ must be filled with entities of type $t$, the following will be true: $(p\ i)$, $(=\ (s\ i)\ j)$, and $(t\ (s\ i))$. These are the only facts we will be using in our discussion of ropes and hangings.

### 11.2.2 Bayesian networks

Bayesian networks are a way of representing information about probability distributions, and are particularly convenient for representing networks of causal influences. The idea is that some set of random variables (for us, boolean variables, denoting partial states of affairs, existence of actions, objects, etc.) influence other random variables. We write the variables as nodes, and the influences as directed arcs, with the direction indicating causality (or picked for convenience in non-causal situations). Figure 1 shows a piece of a medical diagnosis network. (In "right heart syndrome" the heart has trouble pumping blood, and some backs up into the lungs.) Bayesian networks are particularly useful because they conveniently summarize independence.

We will not attempt to defend our use of probability theory in plan recognition other than to note that like all abductive tasks, plan recognition requires putting together many pieces of evidence into a probable, but not certain, conclusion. Probabilities are an obvious candidate

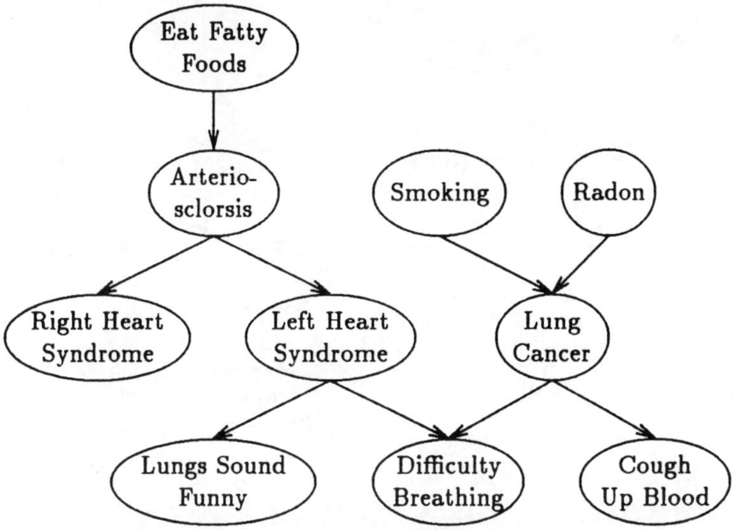


**Figure 1:** Bayesian network for medical diagnosis

for such a task. Nor will we defend our choice of Bayes networks as our probability formalism. Finally, we will not address the issue of the tractability of computing probability distributions in these networks. In general the problem is intractable. If one has a singly connected network (at most one undirected path between any two nodes) there is a polynomial algorithm due to Pearl and Kim [Pearl, 1988]. We expect that the networks we need will, in general, be multiply connected, and we are looking for ways to approximate the probability distribution.

## 11.3 The "Knob" Theory

Now we combine our first-order formalism for plans and their associated objects with Bayesian networks. Figure 2 gives our Bayes network reconstruction of the "rope in the closet" example, though to keep things simple it ignores Jack's role, just concentrating on the clues provided by the words "kill" and "rope." The link between (kill k1) and "kill" represents the idea that the presence of a killing (in the story) caused

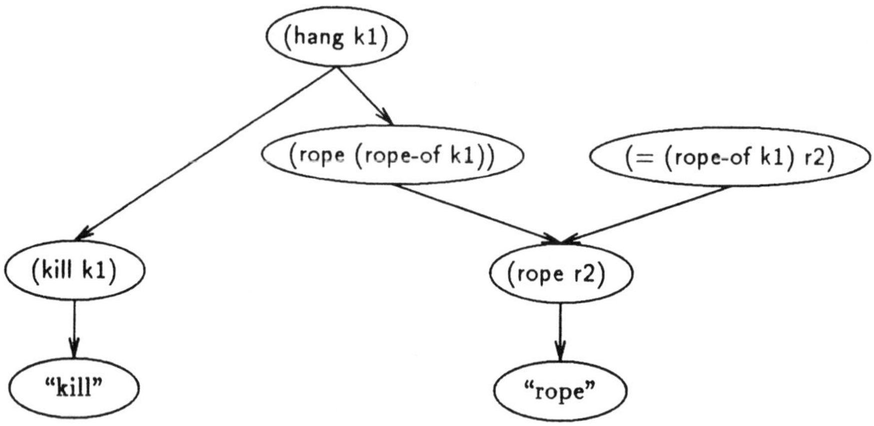


**Figure 2:** Partial network for the "rope in the closet" example

the use of the word "kill."[1] Figure 2 also represents the connection between the hanging event and the killing, and the connection between the presence of a rope and the word "rope." The most interesting part, however, is the connection between the hanging event, and the rope. As in our first-order theory, Figure 2 says that the presence of the hanging implies the presence of *some* rope, which is denoted by the term (rope-of k1). However, this might or might not be the same as the one in the story r2. Thus, the presence of a hanging event will only predict the presence of r2 if the two are the same, (i.e., (= (rope-of k1) r2)). Thus both of these condition our expectation that r2 is a rope. In particular, if both hold then (rope r2) holds with probability 1.

To complete Figure 2 we also need probabilities (the prior probabilities on the topmost nodes, and conditional probabilities on child nodes

[1]Some people have trouble with the causal interpretation of the links between states and events in the world, and the words in the text, as when we said that the killing event caused the use of the word "kill." The way to think about this is that we have an observer who is writing down what he or she observed. Thus the observation of a killing caused her to write it down, and thus use this word. Obviously the real story is much more complex since it is impossible to write down everything one observes. Thus our causal connection is probabilistic, reflecting the fact that we do not have all of the information and causal influences specified (and probably never could). The same is true in the medical diagnosis case.

given the parent nodes). The particular probability which will prove most important for our discussion is the prior for (= (rope-of k1) r2). Before we can say much about it, however, we must figure out what such a probabilistic node *means*. In [Charniak and Goldman, 1989]. we discuss this issue. The key point here is that terms like k1 and r2 are to be interpreted as the outcomes of experiments, where the sample space is all the things in our domain: objects, actions, whatever. In other words, these are arbitrary symbols and could, in principle denote anything. Of course, since we postulate that they are the referents of words in our text, and since the words only can refer to certain kinds of things (the word "rope" cannot refer to a tomato, or at least the probability of this is *very* low) the words used start the process of pinning down exactly what the terms denote. Similarly, n-ary function symbols (e.g., rope-of) denote functions from n members of the sample space to a member of the sample space. The point here is that until we know more, the probability of (= (rope-of k1) r2) is *very* low. It is the probability that two arbitrary entities will turn out to be the same. If you imagined a world populated with, say $10^{20}$ entities, then this would be on the order of $10^{-20}$. Also note that any equality statement will have this same prior probability, and thus we are not talking about something particular to this example, but rather a "fundamental constant" of the system, henceforth denoted $E$. It says, in effect, how likely things are to be associated. (Remember, in our first-order frame representation, associating one entity with another is having one fill a slot in another, and this is expressed using equality).

[Charniak and Goldman, 1989] also discusses the value of $E$, and points out that in realistic stories and realistic observations there is temporal and spatial locality, in the sense that objects observed in sequence will come (typically) from the same portion of "space-time". This raises the probability that two things will be equal, because when restricted to a small part of space-time there are fewer different objects around. In [Charniak and Goldman, 1989] we claim that when this constraint is taken into account, networks for our prototypical stories give reasonable "life-like" probabilities. That is, in the first the probability of hang is low, in the second it is high.[2] For this chapter the

[2]Those who do consult [Charniak and Goldmann, 1989] will note that the net-

reader will have to take this on trust. In what follows we will assume that the spatial/temporal locality constraint has already been taken into account.

We discuss this particular "fundamental constant" ($E$) because changing it will cause the network to assign values to the probability of (hang k1) which range from the story-like, on one hand, to life-like on the other. To see why this is the case it is necessary to have some feel for the flow of probabilities in the network. Setting "rope" and "kill" to 1 causes (rope r2) and (kill k1) to now have probability 1. The latter will push the probability of (hang k1) to a number like, say, $10^{-3}$. This would be saying that one out of a thousand killings is a hanging. The question then becomes, will the evidence from (rope r2) cause it to go even higher. If $E$ is very low (that is, lower than the prior probability of something being a rope) the answer is no. In essence, given this low value, it is more likely that a rope appeared in the story "by accident" than it's presence be dependent on this equality statement. But as we raise the value of $E$ it becomes easier and easier to explain (rope r2) on the basis of assuming both (rope (rope-of k1)) and (= (rope-of k1) r2). This then raises the probability of (rope (rope-of k1)), and in turn (hang k1). More formally, let $e$ denote (= (rope-of k1) r2), $k$ denote (rope (rope-of k1)), and $r$ denote (rope r2). Then $P(r \mid e, k) = 1$, $P(r \mid e, \neg k) = 0$, and $P(r \mid \neg e, k) \approx P(r \mid \neg e, \neg k) = P(r)$. If we assume, as will be the case here, that $P(r) \ll 1$ and $P(k) \ll 1$, then at the point where $P(e) = P(r)$, $\frac{P(k|r)}{P(k)} = 2$. (The proof just takes $P(k \mid r)$, expands it using Bayes theorem, and expands out terms involving $r$ using the above conditional probabilities along with the independence assumptions implicit in Figure 2.) In other words, by setting $E$ to be the prior probability that something is a rope we will double the posterior probability that (rope (rope-of k1)). This will have a similar effect on the probability of hanging. Increasing $E$ still further will have an approximately linear effect of the probability of hanging until the latter starts to approach one. Thus at high values of $E$ we get story-like interpretations, while when $E$ is at or below $P(r)$ we get "life-like"

works there differ from the ones here in the placement of the equality node. Nothing in either theory depends on this difference. Rather differing pedagogical needs dictated the difference.

interpretations.[3]

This then gives us our first, crude, theory for combining both "story-like" and "life-like" plan recognition in a single system. By adopting the Bayesian framework given here the parameter $E$ becomes a "knob" which can be adjusted to give the kind of system we want, a story reader or a real-life plan recognizer.

## 11.4 The "Mention" Theory

The "knob" theory shows that it is possible to have a unified theory, but it does not explain why it is that we needed two theories at all. Why is it that plan recognition differs in stories and life? The knob theory takes it for granted that we need two versions of plan recognition and just shows how to model them in a single system.

We do have some intuitions about why plan recognition is different in stories. As we said in the introduction, in stories things are expected to be relevant, in life they are not. Or to put this another way, we have beliefs about what authors choose to include in a text and what they leave out. They include the relevant, interesting things, and delete the rest. This suggests that we progress toward an explanatory theory by including more of the factors which lead to things being put into the text.

We start by looking more closely at why a writer uses a certain word (e.g., "bank") in a text. In our very simplified model the writer uses a word because the object in question, first-national-43, is of a particular type, savings-institution, and a word of English, "bank," is used to denote such objects. While this causal connection is necessary, it is not sufficient. In our simplifications we have chosen to ignore another necessary component behind the use of the word "bank," the idea that the writer wanted to mention first-national-43. Let us now complicate our lower portions of the Bayesian network to look like the fragment in Figure 3. Needless to say, as a theory of word-choice,

[3]Note that turning $E$ beyond $P(r)$ will still not necessarily be sufficient to hook up very common things, like, say, leaves, because their prior probability is still higher. This seems reasonable.

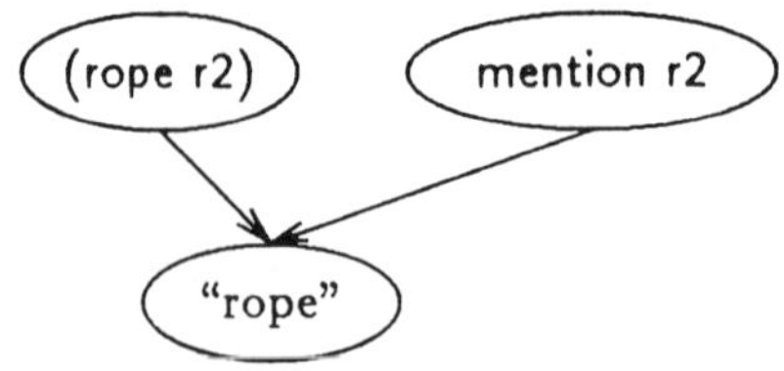


**Figure 3:** Why one uses a particular word.

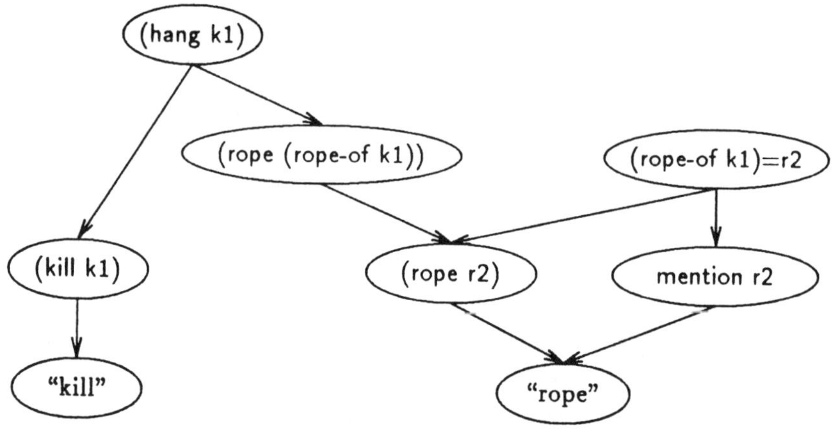


**Figure 4:** Mentioning the rope in the closet

sentence generation, etc., this is still pretty poor, but it is enough for our current purposes.

Next we want to ask why someone will mention a particular entity. This is a complicated matter, and we do not have a theory. Rather we want to show that there is good reason to believe that if we had such a theory it would take over "turning the knob." That is, there would be connections in our Bayesian network between the node (= (rope-of k1) r2) and mention r2. To see this, consider the quite unobjectionable rule, "objects used in already mentioned plans are more worthy of mention that other objects (everything else being equal)." This is not much of a theory of "mention," but it is hardly objectionable. Figure 4 shows the "rope in the closet" network with this rule added in. Note that this rule connects the equality statement with "mention." The reason is, of course, that in our theory we state that an object is used in a plan by asserting an equality between mentioned objects and the plan-role. That is (= (rope-of k1) r2) states that r2 fills the rope-of slot in some plan. Now suppose that objects filling roles in already mentioned plans are $k$ times more likely to be mentioned than an arbitrary object. That is

$$\frac{P(mention|role\ in\ plan)}{P(mention)} = k$$

An application of Bayes formula is sufficient to establish

$$\frac{P(role\ in\ plan|mention)}{P(role\ in\ plan)} = k$$

That is to say, the probability of the equality statement goes up by $k$ given that the word has been mentioned. Thus the fact that the author has mentioned an object is grounds for increasing the (posterior) probabilities of the equality statements. Furthermore, my guess is that $k$ is a very large number, since very few objects in a scene will be mentioned, while a good fraction of those involved in plans will be mentioned. Thus, even an unsophisticated rule like this may well be sufficient to modify the probability of the equality to give us "story like" performance for the class of inferences discussed here. Whether or not this particular "mention" rule is sufficient, this suggests that some such rule will be.

## 11.5 Conclusion

It seems clear that we can expand this theory a great deal without much change. So, while we have discussed inferences about the presence of objects, the theory should work for some inferences about states and events as well. Consider a story like:

> Jack took Mary to a restaurant. The menu was in French.

Lots of people go to fancy restaurants and most of the time they have no trouble reading the menu. However in this story we wonder if there will be a problem. Our rule for objects filling plans is clearly inadequate for this example, but one which discussed causal consequences of states would be reasonable.

Thus we have the beginnings of a theory for why inference in text comprehension is different from real life. It assumes there is a single parameter which governs the difference (for the general case there could well be a small set of parameters) and that assumptions about why authors include facts in texts modify the posterior probability of the parameter(s) in stories. For those (like the authors), who believe that plan recognition in life must be evolutionarily prior with the story process parasitic off of it, the fact that this theory takes the real-life case as basic is also satisfying.

## References

Charniak, E., and Goldman, R.P., [1989]: "A semantics for probabilistic quantifier-free first-order languages, with particular application to story understanding," *Submitted to IJCAI-89.*

Kaplan, S.J., [1978]: "Indirect responses to loaded questions," *Theoretical Issucs in Natural Language Processing - 2.*

Pearl, J., [1988]: *Probabilistic Reasoning in Intelligent Systems: Networks of Plausible Inference*, Morgan Kaufmann, San Mateo.

# Chapter 12

# New Foundations for Qualitative Physics

Jean Petitot* and Barry Smith†
**Ecole des Hautes Etudes en Sciences Sociales, 54 Bd. Raspail, 75270 Paris, France*
†*Internationale Akademie für Philosophie, Obergass 75, 9494 Schaan, Liechtenstein*

## 12.1 Preamble

Physical reality is all the reality we have, and so physical theory in the standard sense is all the ontology we need. This, at least, was an assumption taken almost universally for granted by the advocates of exact philosophy for much of the present century. Every event, it was held, is a physical event, and all structure in reality is physical structure. The grip of this assumption has perhaps been gradually weakened in recent years as far as the sciences of mind are concerned. When it comes to the sciences of external reality, however, it continues to hold sway, so that contemporary philosophers - even while devoting vast amounts of attention to the *language* we use in describing the world of everyday experience – still refuse to see this world as being itself a proper object of theoretical concern.

Here, however, we shall argue that the usual conception of physical reality as constituting a unique bedrock of objectivity reflects a rather archaic view as to the nature of physics itself and is in fact incompatible with the development of the discipline since Newton. More specifically, we shall seek to show that the world of qualitative structures, for example of colour and sound, or the commonsense world of coloured and sounding things, can be treated scientifically (ontologically) on its own terms, and that such a treatment can help us better to understand the structures both of physical reality and of cognition.

A number of recent moves have been made by workers in the field of artificial intelligence in the direction of a theoretical account of the qualitative level of objective reality. We can point, for example, to the idea of a 'naive physics' as this has been propagated by Patrick Hayes, [Hayes, 1985] and to the qualitative physics of Kleer & Brown, [1984]. Parallel ideas are present also in the project of a 'semiophysics' - a physics of the salient structures in reality – that has been advanced by the French mathematician René Thom. Thom's ideas are propounded, interestingly enough, in the form of a commentary on Aristotle's *Physics* [Thom, 1988]. For it was not always the case that philosophers were disposed to cast aspersions on the project of a science of the qualitative world. To Aristotle and his disciples physics itself was indeed a qualitative discipline, and modern day practitioners in the field of naive physics have recognized that there are valuable insights to be gained from the work of medieval thinkers, such as Buridan and Oresme, still working within a broadly Aristotelian framework.[1] Thomas Reid and other Scottish common sense philosophers can likewise be seen as having explored the world of qualitative reality in ways relevant to more recent experiments. In the writings of thinkers such as Reid, however, as also in the work of the medievals, the issue is for obvious reasons not addressed as to the proper relation between the (qualitative) description of commonsensical reality and physical science in the modern (quantitative) sense.

The question thus arises as to who, in the philosophical tradition, was the first exponent of what might be called a *sophisticated naive physics*, which is to say a theory of the commonsensical domain whose relations to physics proper are made the subject of explicit theoretical

[1]See e. g. [Holland et al., 1986] p. 208.

concern. Claims might be made in this respect for Whitehead, whose "On Mathematical Concepts of the Material World" [Whitehead, 1906] stands at the beginning of a long and valuable tradition of formal ontology embracing also, *inter alia*, the work of J. H. Woodger.[2] It seems, however, to have been Husserl's *Crisis of European Sciences* of 1936 which first addressed in explicit fashion the relation between the ontology of the commonsense world – called by Husserl the 'theory of the structures of the life-world' – and post-Galilean physics.[3] And Husserl's ideas as presented both in this work and also in his earlier writings on formal ontology[4] will surely be recognized by future researchers in the area of naive physics as one crucial philosophical pillar of their discipline.

It might, for a number of reasons, seem somewhat incongruous to run together such diverse intellectual currents under the single umbrella of what we are still somewhat loosely calling 'naive' or 'qualitative' physics. There is, first of all, an important divide between those, like Thom, who are concerned to develop the physics of salience as a mathematical discipline, and those, like Husserl, who see the structures of the life-world as demanding a theoretical treatment of a quite different sort.[5] There is a deep divide also between those thinkers – such as Aristotle – who see the discipline of naive physics as a science with its own distinctive subject-matter, and those – like most contemporary workers in the field – who see naive physics in quite other (cognitive or psychological) terms. Of course, given the assumption mentioned at the start of this chapter, it is not difficult to see why the first alternative should nowadays prove so unpopular. If reality *an sich* is conceived as being captured exclusively and exhaustively in the (suitably perfected) equations of a purportedly monolithic discipline of standard physics, then there would seem to be no room for any additional science of the structures of commonsensical reality – unless, that is, such a science should be a sort of psychology in disguise, a science of 'knowledge simulations' or of 'mental models', readily associable with investigations

[2]See e.g. [Woodger, 1937] and [Woodger, 1939].

[3]See [Husserl, 1936].

[4]These works and their influence are treated at length in [Smith, 1982].

[5]See [Petitot, 1989b].

in, for example, the sphere of 'children's physics'.[6] There is a danger, however, that the idea of a science of commonsensical reality will in this way be confused with the quite different and patently absurd idea of a 'commonsensical science' – a confusion of the sort which seems to lie at the heart of discussions of 'folk psychology' by Churchland and others.[7] Folk psychology is of course not a science, but a matter of sheer popular prejudice. From this, however, we clearly cannot conclude that there cannot be a sophisticated science of mind, and nor can we conclude from the muddled state of many folk beliefs about the commonsensical world that a sophisticated science of the structures of (the objective component of) this world is ruled out *a priori.*

Workers in the field of artificial intelligence may be able to afford to ignore such issues and concentrate on the practical job of simulating relevant human beliefs and processes of reasoning in formal theories, irrespective of the issue as to whether the propositions which thereby result are true or false of any independent reality. Thus they may take the view that all that matters, from their practical point of view, is merely the extent to which one obtains desired end-results in the sphere of automated reasoning. Neither the psychological nor the pragmatic conception of naive physics can be ultimately satisfying on their own terms however. For both leave open the question why it is that just these mental models or systems of beliefs should have arisen as they did and why they should have survived so long. Moreover, they leave open the question as to why it is that they should have the power to sustain so remarkable a facility of both thought and action. In order to answer these questions one must, it seems, adopt a wider theoretical focus, taking account of the structures of the world in which such thought and action is realized. One needs, that is, to place one's theories or simulations of the psychology of human thought processes within the wider framework of an ontology. One very tempting hypothesis then consists in the idea that the remarkable facility which humans manifest in thinking and acting on the level of everyday experience is made explicable, at least in part, precisely by the existence of corresponding stable structures on the side of reality.

[6]See [Holland, et al., 1986] pp. 206ff.

[7]See e.g. [Churchland, 1979].

It is this hypothesis – a hypothesis which comes down in the end to the view that there is a level of reality which enjoys a certain sort of intrinsic intelligibility – which we shall pursue in what follows. Thus we shall be concerned to establish (the foundations of) a theory of the qualitative or commonsensical world conceived as a relatively autonomous level of reality confronting us in everyday experience.[8] Such a theory must rest on one or other form of Aristotelian ontology, in the sense of an ontology recognizing enduring animate and inanimate substances manifesting an opposition between form and matter, possessing sensible and non-sensible qualities and undergoing changes (events and processes) of various sorts. On the appropriateness of such an ontology there is wide agreement among all practitioners, whether they work in the field of philosophy or in artificial intelligence, and whether or not they are willing to give credence to the corresponding propositions as propositions which are true of some independent reality. Thus it is remarkable to observe the extent to which Hayes' list of conceptual 'clusters' or sub-theories of the discipline of naive physics as he conceives it,[9] corresponds to the original master list of categories supplied by Aristotle.

The Aristotelian ontology must in addition recognize species and genera (or "natural kinds") which these entities, both substances and their accidents, instantiate, and it must recognize further that the instances in each kind are divided into circles of more and less standard or prototypical instances. The prototypical instances in each species can then be expected to be more readily discriminable (salient, *prägnant*) than their non-prototypical counterparts, and also more readily able to give rise to correspondingly skilled responses on the parts of perceiving and acting subjects. All of these features were investigated extensively by successive generations of philosophers inspired by Aristotle – up to

[8] We are not, for the moment, interested in the precise relation between "qualitative" and "commonsensical". Suffice it to say that the qualitative domain as specified below extends more widely than does the domain of commonsensical experience, for example in including colour-like qualities off the visible spectrum. On the other hand the world of commonsensical experience embraces dimensions of ontological form, above all the dimension of *substance*, which are skew to the strictly qualitative sphere. See [Smith, in preparation].

[9] See [Hayes, 1985] pp. 18ff.

and including Husserl. The Aristotelian qualitative ontology was however called into question by Galileo and his successors. Above all, substances and sensible ('secondary') qualities came to be eliminated from the view of the world accepted by the physicists, along with the whole apparatus of form and matter, natural kinds, prototypical instances, and so on.

Clearly, though, the qualitative or commonsensical ontology can be Aristotelian only in a broad sense. Thus the space of this ontology must be three-dimensional and global in type, as contrasted with the purely local space of Aristotle. Substances occupy volumes of this space and move continuously through it. They have closed spatial boundaries which delimit and separate them from other substances and they are capable of communicating impetus. And, most importantly for our present purposes, the sensible qualities inhering in such substances will manifest *qualitative discontinuities* which may or may not coincide with the boundaries which mark their exterior surfaces in space. Consider, for example, the case of a black dog with brown spots. Here, two sorts of qualitative discontinuities can be distinguished. On the one hand are the discontinuities corresponding to the exterior apparent contours of the dog; and on the other hand are the internal discontinuities on the surface of the dog (for example the contours of the spots on his back).

## 12.2 Manifestations of Matter: I. Spatial Movement

If, as we claimed, modern physics is not the science of some ultimate bedrock of reality, what, then, is it? Crudely speaking, it is a science which deals with a rather limited number of ways in which matter manifests itself in qualitative reality. It deals with these manifestations not, however, as denizens of the qualitative world but as it were in purified form, as quantities or magnitudes. Its purchase is, in other words, exclusively mathematical, and it seeks to use mathematical devices to *explain* the given manifestations by showing how they are subject to formal laws or principles. Qualitative reality is here, in a way, preserved; but it comes to be filtered entirely through structures of a formal or quantitative sort.

Classical mechanics, for example, seeks to explain in a mathematical way and in a single, unified framework all the diverse expressions of that manifest property of matter which is *spatial movement*, from the movements of pendulums and the orbits of planets to turbulence in fluid dynamics and the thermodynamic phenomena captured by statistical mechanics. Such movements are represented within the theories of mechanical dynamics using entities such as vectors (in the case of velocities, accelerations, gradients, etc.), or tensors (of impulsion, of deformations in continuous media, etc.), or differential forms (of flux, divergence, curl, etc.).

Vectors, tensors and differential forms, now, are all such as to possess intrinsic geometrical meanings, in the sense that they can be understood as entities having determinate structures that are independent of the co-ordinate apparatus we use to describe them. One consequence of this, is that the descriptions of movement yielded by classical mechanics can and must be true independently of whatever we happen to choose as a co-ordinate frame. This is an *a priori* (pre-physical) requirement on the descriptions of the theory. It implies, first of all, that the differential entities of which the theory treats must enjoy the specific mathematical property that they vary covariantly in respect to the Galilean relativity group. Because no point in time is distinguishable physically from any other, it is impossible physically to determine an absolute origin of time: with this fact is associated the relativity sub-group of time-translations. Similarly, it is impossible physically to determine an absolute origin for the co-ordinates of space, or an absolute direction in space, and with these facts are associated respectively the relativity sub-groups of spatial translations and spatial rotations.

The Galilean group is a group of *symmetries* of space and time: every point in space or time is indistinguishable from every other. In general, the symmetries of a theory express what it is impossible to know absolutely according to the theory. Because it is impossible physically to determine an absolute point in time or an absolute point or direction in space, we are forced to fix these entities conventionally: we choose arbitrarily a zero for time, a zero for space, a co-ordinate frame, units of measure, and so on, and it is precisely because these entities are not physical that our physical descriptions must be independent of the given choices.

It turns out, however, that this seemingly trivial *a priori* constraint of co-variance has important physical consequences. A mechanical system is completely described by a certain function, called the Lagrangian, related to its energy. One of the greatest theorems of classical mechanics, namely Noether's theorem (which can be generalized to physical theories of many other sorts), says that, if the Lagrangian is invariant through the transformations of any given one-dimensional sub-group of relativities, then there is a certain physical quantity correlated therewith, which is *conserved* through every movement of the system.

The exact form of Noether's theorem is that to every one-parameter group of symmetries of the Lagrangian there is correlated a law of conservation of a physical quantity. Time translations are correlated in this way with the law of conservation of energy. Spatial translations are correlated with the law of conservation of momentum. And spatial rotations are correlated with the law of conservation of angular momentum.

From correlations of this type, which are valid *a priori*, we can derive an enormous amount of physical description – to the extent that many physicists will readily assert that the whole physical content of classical mechanics is exhausted by such laws of conservation. Einstein's celebrated law of the equivalence of mass-energy is itself a direct consequence of Noether's theorem applied to the new relativity group of four-dimensional Minkowskian space-time.

## 12.3 Manifestations of Matter. II: Quantum Phenomena

Matter manifests itself phenomenally not merely via movement but also for example via *spectral rays*. Quantum physics can be seen precisely as the physics which relates to this mode of manifestation of matter as classical mechanics relates to movement. In addition to 'external' space-time, with its Galilean or Einsteinian relativity group, quantum mechanics relates to what are called 'internal' quantum numbers. These are new physical quantities which characterize elementary particles (electric charge, isospin, charm, colour). And here again there are certain *a priori* pre-physical constraints which prove to be of fundamental significance in determining the nature of the objects of the physical theory which results. For example it turns out empirically that in the nexus of strong nuclear interactions the proton and the neutron are indiscernible. The symmetry between the two is called the isospin symmetry. Applying Noether's theorem we get a conservation law which is the law of conservation of isospin in nuclear reactions.

Another example of this same phenomenon turns on the fact that it is impossible, within a single quantum system, to individuate by physical means an elementary particle in a group of elementary particles of the same type (for example one electron in an atom with many electrons). This fact, again, seems at first not to have much physical content, but to reflect only an inadequacy on our part, an *a priori* indiscernibility. The Lagrangian here becomes an operator which operates on a function describing the quantum state of the system. This Lagrangian is invariant with respect to the symmetry that is represented by the group of permutations of the particles within the system. Two alternatives turn out to arise here. In some cases permuting the particles leads to no change in the function: the function is symmetric. In other cases the function is antisymmetric and leads to a change of sign for an elementary permutation (a transposition). This opposition could hardly be more important. It generates those physical properties of matter which are known in quantum mechanics as the correlation between spin and statistic. Those systems which are antisymmetric are constituted by particles – called 'fermions' – which have a half-integral spin (1/2, 3/2, 5/2, etc.). Such particles, which are particles of matter, are subject to Pauli's exclusion principle, which states that two fermions in the same position in space-time cannot have the same quantum numbers. It is this principle which explains, for example, why all the electrons of an atom must have different systems of quantum numbers (electrons are fermions). This explains in turn why one needs orbits of electrons, why matter does not collapse, and therefore also why matter exists macroscopically and so manifests chemical properties. Those systems, on the other hand, which are symmetric, are constituted by particles of integral spin. These particles, called bosons, are not particles of matter, but particles vehiculating interactions between particles of matter. For example it is well known that the photon is the particle vehiculating electromagnetic interaction. For the bosons Pauli's Exclusion Principle is not valid. Thus we can have superposition of bosons in the same position in space-time – which explains such physical phenomena as lasers and superfluidity. Here also, then, we have deep properties of matter, which are in fact the physical translation of certain *a priori* constraints.

## 12.4 Manifestations of Matter. III: Qualitative Discontinuities

An "observable" in modern physics must be *measurable*, in conformity with the move effected already in the work of Galileo from qualitative to quantitative aspects of reality. But for something to be measurable, there has to be the possibility of invariance. Thus without the possibility of constant temperature, temperature could not be a physical magnitude. A physical magnitude presupposes the possibility of conservation in certain ideal conditions. But now again, this means that the required stability must be imposed as an *a priori* condition on the possibility of theory, and the sort of stability we impose will determine the sort of theory we end up with. This reinforces the correlation between *a priori* constraints and what the objects of the theory are.

We can now see clearly why physics of the post-Galilean sort cannot serve as the description of some autonomous bedrock of reality. Post-Galilean physics involves, if one will, an ineliminable (or irredeemable) Kantian dimension. It yields not a picture or description of some theory-independent reality, but a certain unified system of *quantities*, of mathematical reconstructions of regularities in the manifestations of matter, reconstructions that are in no small part dependent upon *a priori* constraints which must be satisfied if the given quantities are to be graspable at all. The world of standard physics is in this sense a human construction.

Our specific problem, now, is that of establishing how physical theories can be enriched sufficiently to capture in scientific fashion the features specific to qualitative reality. The ground of a solution is in fact already to hand: physics, for all that it is restricted to the quantitative, does indeed deal with just the manifestations of matter from out of which the qualitative world is composed. What it does not do is to deal with those very special sorts of ways in which manifestations of matter are composed or knitted together which are relevant to the world of qualitative experience. Can we, then, build up a science of the properly qualitative modes of manifestation of matter? Can we, in other words, find a way of bridging the gap between quantity and quality, or between the physical and the qualitative modes of manifestation of matter?

Here the obvious suggestion is that the qualities manifested locally in qualitative reality be represented as degrees of appropriate intensive magnitudes: colours via frequencies, qualities of hot and cold via temperatures, etc. Such representations will most importantly preserve the spatial or temporal *variations* in the represented qualities, and it seems reasonable to suppose that it is in such variations that the relevant qualitative information will be contained. *Only some types of physical variation will however be able to represent variations of the qualitative sort.* Simple mechanical systems (pendulums, for example) fall entirely out of court in this regard. How, then, are we to focus in on variations of the relevant types? Here the key idea is due to René Thom. Variations in intensive magnitudes will, clearly, be either continuous or discontinuous. Thom's idea is that the science we require should take as its primitive *qualitative discontinuity*, which is to say discontinuous variations in qualities as represented quantitatively in the given fashion. The theory which results would then be a science of those manifestations of matter which are associated with discontinuities of variation in intensive magnitudes, just as classical mechanics is a science of those manifestations of matter we call spatial movement.

The steps involved in building up a science of qualitative reality along these lines can here be sketched only in broad terms:

(i) We must first of all convince ourselves that the given primitive truly does yield a science of qualitative reality;

(ii) We must give an adequate mathematical expression of the idea of qualitative discontinuity.

(iii) We must give an account of how we can move from standard physics to a theory of the qualitative world of the sort desired.

We shall deal very briefly with each of these in turn.[10]

**Ad i:** How do things appear in qualitative reality? For present purposes we may confine our attentions to sensible phenomena. We begin by drawing attention to three characteristic features of the ways things, events, etc. appear in sensation – features first picked out by Husserl:

1. Whatever appears, appears always from one side, presents one (continuously varying) face or aspect, and is correspondingly foreshortened or 'adumbrated'. The science of commonsensical reality thereby divides into one branch which deals with the different sorts of substances, qualities, etc., as they are in themselves, and a second branch which deals with the corresponding types of adumbrations and with the relations between systems or chains of adumbrations and the structures which they are adumbrations of.

2. Whatever appears, appears in such a way as to manifest a foreground-background structure (there is an ever-present difference between topic or focus and the context against which this topic or focus is set into relief).

3. Whatever appears, appears in the context of a spatio-temporally extended whole. There is, in other words, a relation of foundation between sensible qualities and spatio-temporal extension (no colour can, as a matter of necessity, exist without spatial extension, no sound without duration, etc.).

Take, again, our example of the black dog with brown spots as this appears in visual perception. Here we can distinguish first of all the apparent contour of the form or shape of the dog: the exterior boundary of the dog as this appears to us at any given instant (and the fact that we can only perceive one apparent contour at a time is a typical example of foreshortening). Like the aspects of which they form a part, such apparent contours will typically vary continuously (for example as the dog trots off into the night), and it is an interesting problem

[10]For more details see [Petitot, 1989a, b, c and d].

to understand the relation between the three-dimensional boundary of the dog itself and the family of its two-dimensional apparent contours.

The perceived qualities of the dog, now, have a certain spatial extension, and it is in this context that the notion of qualitative discontinuity arises. The qualities distributed across any given spatial extent are either fused phenomenally in the sense that there is no observable separation between them (there is a smooth transition from one colour to another), or they are phenomenally 'separated', where there is discontinuous variation. (Of course something similar arises in relation to auditory qualities and their temporal duration.)

Here there is no intrinsic difference between the apparent contours which correspond to the perceived exterior of the dog and mark the distinction between foreground and background, and the perceived boundaries within the interior of the dog, for example the boundaries separating spots from their surroundings. In relation to either sort of case we arrive at the conclusion that a sensible phenomenon is set into relief in relation to other phenomena precisely where a discontinuity of the given sort has been created by the qualitative moments which fill its extension. Only thus can the phenomenon 'make itself count on its own and be noticed (stand forth for itself)'.[11] It is *separation*, in other words, which accounts for *salience*. Thus we have strong grounds for supposing that qualitative discontinuities can indeed serve as an organizing principle of the qualitative world of the sort we have been seeking.

**Ad ii:** In giving an appropriate mathematical expression to the notion of qualitative discontinuity we follow the topological approach outlined by Thom. Suppose that $W$ is the spatio-temporal extension of a given phenomenon. As a portion of space time, $W$ is of course a topological space with the usual topology. Suppose further that the different qualities which fill $W$ are expressed by degrees of $n$ distinct intensive magnitudes $q_1, q_2, ..., q_n$ , each a function of points $w \in W$. The $q_i(w)$ are then sensible qualities (colour, texture, temperature, reflectance,

[11]Cf. §8 (A239, B244) of Husserl's 3rd Logical Investigation. The ideas in this work, which reveal a remarkable topological sophistication, have unfortunately remained unexploited by Husserl's successors. See [Husserl, 1900/01], [Smith, ed., 1982] and [Petitot 1989b].

etc.), but considered physically,[12] as immanent to the objects themselves and as associated with a certain possibility of measuring.

A point $w$ is called *regular* if all $q_i(w)$ are continuous in a neighbourhood of $w$. Let $R$ be the set of regular points of $W$. $R$ contains a neighbourhood of every one of its points, and hence it is an open set of $W$. Let $K$ be the complementary set of $R$ relative to $W$. $K$ is the closed set of what we can call the non-regular or *singular* points of $W$. Clearly, $w$ is a singular point if and only if there is at least one quality $q_i$ which is discontinuous at $w$. We shall call $K$ the underlying morphology of the phenomenon that fills $W$. $K$ is the system of qualitative discontinuities which sets this phenomenon into relief and makes it salient as a phenomenon. Consider, for example, the morphological organization of a leaf, of a crack in a window pane, of a dog, or of the photograph of a dog.

**Ad iii:** In order, now, to accord physical content to the definition, we must find some way to conceive a morphology $K$ as a manifestation of physical properties internal to whatever underlies or causes the phenomenon in question. Condensing a lot of physical detail into a small space,[13] we can say that the instantaneous states of a physical system *qua* physical are, when taken individually, *transient*: they are too fleeting to be observable. In general, however, there arise effectively observable states of a system: this occurs when trajectories manifest a well defined asymptotic behaviour. Such effectively observable states, the states into which the system repeatedly falls, or into which it tends to fall, are for obvious reasons called the *attractors* of the system. Consider, for example, an oscillating electric circuit. From any initial state the system after some time reaches the stable oscillatory state and so its trajectory is attracted by this state.

[12]Of course certain simplifications are involved here. Thus colour might be accounted for in terms of three distinct qualities: hue, saturation and brightness. Similarly there is no single property of reflectance, which is a macroscopic approximation of a more fine-grained system on the quantum level of a range of properties having to do with the emission-absorption spectra of the atoms constituting the substrate.

[13]Some of this detail is provided in our paper "Physics and the Phenomenal World", in preparation. See also [Petitot, 1989b and d].

Return, now, to a phenomenon having substrate $S$, spatio-temporal extension $W$ and morphology $K$. Choose a non-singular point $w \in W$. The internal state of the substrate $S$ at $w$ can be physically described in terms of some attractor $A_w$. And the $q_i(w)$ are intensive quantities associated with $A_w$. To explain the qualitative discontinuities of the $q_i(w)$, we now let $w_o \in K$ be some singular point of $W$. The idea is that in moving through points $w \in W$ the attractor $A_w$ becomes unstable when we cross $w_o$ and is replaced by another attractor $B_w$.[14] Similarly we can explain what happens at the exterior apparent contours of things by saying that when we cross such a contour the attractor $A_w$ disappears entirely.

In macroscopical physics, there are many examples of such phenomena of transition of the internal state of a system. They are known as *critical phenomena.* A typical example is that of phase transitions in thermodynamics, where a system undergoes a sudden change of thermodynamical phase (for example from solid to liquid or from liquid to gas, from a magnetic to a non-magnetic phase). Such changes occur when a parameter such as temperature crosses a critical value. In this example the qualities $q_i(w)$ are the thermodynamical phases. The behaviour of the underlying physical system is unobservable. What we experience as salient, and what we possess words to describe, is the qualitative discontinuity which is the phase transition.

There are many other critical phenomena: for example shock waves in acoustics, transition to turbulence in hydrodynamics, buckling in elasticity theory, etc. All such phenomena are salient in our perceptual experience. They and their like are the physical support of the qualitative organization of the world of everyday experience.

[14]When the local physical state of the substrate S at $w$ is defined by a dynamical system, one speaks of a *bifurcation* event. When it is characterised by a certain type of symmetry, one speaks of a *symmetry breaking* event.

## 12.5 A Theory of the Commonsense World

We claim that one can succeed in this way in solving the problem (in principle, at least) of relating physics and the qualitative world from the point of view of the mathematics of morphologies. Of course our treatment of qualitative discontinuities, too, does not yield a description of the qualitative world as some monolithic bedrock of reality. In this sense it is at the same distance from an ontology in the strict sense as are classical mechanics, quantum mechanics, etc. For our theory deals after all not with objects (qualities, etc.) in the world, but rather with products of mathematical reconstruction. The difference, however, is that this reconstruction turns out to allow on its own terms a mimicking of just those central features of the Aristotelian commonsensical ontology that were so fatefully abandoned by Galileo and his successors. That is, it offers not only a theory of qualities, but also, in the long run, a theory of substance, of change or process, of typicality, species and categorization, and so on – or in other words an entire theory of the commonsensical world.

But does this theory constitute a science in the strict and proper sense? Certainly it is not predictive in the usual (causal) sense; but then the aim of the qualitative ontology is not the aim of standard physics. The approach does lead to prediction, but only in the sense that it leads to the possibility of our explicating mathematical constraints for different sorts of empirical morphologies. This is a 'prediction' of exactly the same sort as the predictions to the effect that if you have, for example, a crystal, or the envelope of a virus, or a snowflake, or honeycomb, or an ornamentation of the Alhambra of Granada, then the symmetry of the structure is necessarily one of the 'Platonic' symmetries which is allowed by geometry. There exist theorems which make the same type of structural predictions for the possible morphologies $K$. These predictions can be interpreted as abstract mathematical constraints upon the universe of morphological phenomena.

## 12.6 Qualitative Ontology and the Science of Cognition

We have seen how a theory of the commonsense world can be rooted in the physics of the material substrates. But in order to have a plausible theory of the qualitative world we clearly need in addition a psychological-cognitive theory of perception and an account of the link between this theory and the substrate theory, or between perceiver and object of perception. How is the perceiving subject involved in the perceptual explication and cognitive interpretation of the qualitative structure of the commonsensical world? As far as qualities such as colour are concerned, we already dispose of considerable work on these problems and we know something about the steps which lead from physics to the mind.

We have first of all, at the micro-level, the absorption-emission spectra of the atoms making up the substrate. At the macro-level we have the reflectance of the object, which gives rise in its turn to transmission of light of certain wavelengths. At the level of the retina, the light excites the cones and the information (pattern of wavelengths) it bears is processed by these transducers, which is to say it is transformed into neuronal information (frequencies of neuron firings codifying the wavelengths). This gets processed further on its way to the visual cortex, where there occurs the registering of a sensible quality of colour. Through all these steps something is preserved, and from our present point of view it is clear that at least part of what is preserved can be very well explained via the concept of qualitative discontinuity. For this is a concept which applies equally to qualities as realized physically and as apprehended in patterns of sensation in the mind. Wave optics explains (in a highly non-trivial manner) how the very special type of information which concerns qualitative discontinuities can come to be encoded in light (that is to say how singularities can be propagated by light). And the theory of visual perception (for example as propounded by David Marr) clarifies the perceptive endowment which allows us as cognitive systems to detect and to process this information. Briefly, we can say that qualitative discontinuities on the side of the object are qualitatively salient for the subject because the concept of qualitative discontinuity is at one and the same time an objective and a subjective concept [Petitot 1989d].

The interest of the point of view of Marr and his successors is that it reconciles two apparently antagonistic approaches: the information-processing point of view, and the ecological point of view of J. J. Gibson, et. al. In the classical cognitivist paradigm (as exemplified by Fodor, Pylyshyn, et. al.), information processing is essentially reduced to the operations of calculation on symbolic mental representations. These operations are essentially syntactic: the cognitivists focus exclusively on algorithms and neuronal implementations thereof; they leave no room for any link between the cognitive system and those objective, stable features of the qualitative world which we have placed at the centre of our concerns. If, however, one wants to introduce such objective structures of the environment into one's account of perception and cognition, as the ecologists do, then one is committed to making such structures constrain the information-processing devices.

Marr explains that what Gibson considered as the 'extraction' (pick-up) of invariants from the environment can be understood as a form of information processing capable of being explicated in computational terms. But a computational theory in Marr's sense must not only focus on algorithms and neuronal implementations; it must in addition comprehend each algorithm in relation to the type of object-derived information which it processes. But then, because the algorithms employed are determined by objective properties of the environment, it follows that we have a means of reconciling the syntactic status of the algorithms with an ecological semantics. Methodological solipsism is hereby overcome, and the world is once more made safe for qualitative experience.

## References

Churchland, P., [1979]: *Scientific Realism and the Plasticity of Mind*, Cambridge University Press. Cambridge.

Hayes, P.J., [1985]: "The Second Naive Physics Manifesto", in Hobbs, J.R., and Moore R.C., (eds.), *Formal Theories of the Commonsense World*, 1-36, Norwood, NJ. Ablex.

Holland, J.H., Holyoak, K.J., Nisbett, R.E., & Thagard, P.R., [1986]:

*Induction. Processes of Inference, Leaning, and Discovery.* MIT Press. Cambridge, Ma. and London.

Husserl, E., [1900/01] *Logical Investigations*, Findlay, J.N., (trans.), Routledge and Keegan Paul, London. 2 vols. 1970.

Husserl, E., [1936]: *The Crisis of European Sciences and Transcendental Phenomenology* Carr, H.D., (trans.), Evanston. Northwestern University Press. 1970.

Kleer, J.D., & Brown, J.S., [1984]: "A Qualitative Physics Based on Confluences", *Artificial Intelligence*, 24, 7-84 and in Hobbs and Moore (eds.), op. cit., 109-183.

Petitot, J., [1989a]: "Morphodynamics and the categorical perception of phonological units", *Theoretical Linguistics*, 15, 1/2, 25-41.

Petitot, J., [1989b]: "Structuralisme et Phénoménologie", in Petitot, J., (ed.), *Logos et Théorie des Catastrophes*, 345-376, Editions Patiño, Genève.

Petitot, J., [1989c]: "Hypothèse localiste, Modèles morphodynamiques et théories cognitives", *Semiotica*, 77, 1/3, 65-119.

Petitot, J., [1989d]: "Forme", *Encyclopaedia Universalis*, XII, 712-728.

Petitot, J., and Smith, B., [in preparation]: "Physics and the Phenomenal World".

Smith, B., (ed.), [1982]: *Parts and Moments. Studies in Logic and Formal Ontology*, Philosophia. Munich.

Smith, B., [in preparation]: "The Structures of the Commonsense World".

Thom, R., [1988]: *Esquisse d'une Sémiophysique. Physique aristotélicienne et Théorie des Catastrophes*, Paris: Intereditions.

Whitehead, A.N., [1906]: "On Mathematical Concepts of the Material World", in *Philosophical Transactions of the Royal Society of London*, series A, vol. 205, 465-525, repr. in Northrop, F.S.C., and Gross, M.W., (eds.), *Alfred North Whitehead. An Anthology*, 7-82, Cambridge University Press. Cambridge 1953.

Woodger, J.H., [1937]: *The Axiomatic Method in Biology*, Cambridge University Press. Cambridge.

Woodger, J.H., [1939]: "The Technique of Theory Construction", in *International Encyclopedia of Unified Science, vol. II, no. 5*, University of Chicago Press. Chicago.

# Chapter 13

# Can't Kant: Smith on Folk Physics

Adam Morton
*Department of Philosophy, University of Bristol, Bristol, England*

## 13.1 Smith

The most profitable line for me to take, I think, is to defend my discussant. This may surprise you, and I fear the reasons for it may annoy Barry. They are that I think the Smith & Petitot chapter has some good and interesting ideas, but they are so hedged about by metaphysical extravagance on the one side and bare unfulfilled promises of relevance to AI on the other, that one very natural reaction would be just to dismiss them. So I shall try to extract some important points from their chapter.

## 13.2 Statement of Faith

To begin, here is a conviction that Smith and I share, and share with many physicists and a number of European philosophers, but not with very many English speaking philosophers. The general purpose of this paper is to state this conviction, in a way that makes it seem interesting and important and possibly true, and to argue for its relevance to issues about naive physics. I shall only gesture towards issues about naive psychology.

The central point - henceforth Central Point - is the existence of mathematical constraints on the variety of forms that the phenomena produced by some kinds of laws can take. The constraints usually have an algebraic or topological basis. Smith cites Noether's theorem but the most dramatic instance is Thom's theorem and the most interesting, but as yet mysterious, is the series of topological results now beginning to underpin the patterns that the "empirical" study of chaotic systems has turned up. What one has in all these cases are classes of physical systems which though governed by common laws show a great variation in their behaviour, a variation which is yet constrained to occur within certain very abstract limits. Although differences of systems or of their initial and boundary conditions can lead to a great variety of histories, there is superimposed on this variety an *a priori* restriction on the "qualitative" forms that are found.

Two examples of this. The clearest is catastrophe theory. I am not going to give an exposition of it. But I do want to make explicit the important point that catastrope theory is not a theory, but an explanatory strategy. That is, it consists of some advice about how to provide explanations of a wide class of phenomena, essentially those in which a system governed by laws which describe continuous changes in its basic quantities exhibits discontinuities in its behaviour. The laws in question can have a variety of forms as long as they can be got into a rough Hamiltonian shape. Then the strategy is to divide the variables into two classes, control parameters and system variables, and apply some topology to the projections from the larger-dimensional surfaces of the system's trajectories conceived of in terms of all its variables to the smaller-dimensional subspace of the control parameters. The point of doing this is that mathematical constraints – Thom's theorem – limit the forms projections of singularities – folds – from the larger space to the smaller one can take, so that one gets in the end a classification of the points at which discontinuites in the system's behaviour arise.

The explanatory strategy is not a theory because it needs to be supplied with a set of dynamic principles from a pre-existing background theory, and because there is no prescription for finding the right splitting of the parameters. It's "just" a cookbook backed up by a theorem.

A cookbook for making models and using them to explain things. In fact, it is very unclear when the modelling results in an explanation. This is a very delicate question: at one extreme are cases when the underlying dynamics can do all the explaining, and the catastrophe theoretical gloss is just a heuristic, and at the other are cases where one does not in fact have any dynamic laws, and there it is not clear that there is any explanation at all being given. (These are the cases that give Zeemanism or Thomisme a bad name.).[1] Where the theory has a real bite is somewhere in between, when there is an underlying dynamic and either it cannot be completely specified or predictions of behaviour cannot be derived explicitly from it.

Next example: chaos. The beginning point here is a very surprising fact. Consider chaotic systems, that is, crudely, systems which depend so sensitively on their initial conditions that although governed by deterministic laws their evolution is in principle unpredictable. Again we think in terms of control parameters and state variables, but also in terms of attractors, the regions of state space that the system evolves towards. The explanatory strategy now is to classify such systems in terms of the topological characteristics of their attractors and the way in which the number of their attractors varies as the values of their control parameters change. Then a remarkable fact emerges: the same patterns recur in very different systems. This fact lies behind what seems to have become the main intellectual tool in understanding chaos. One begins with simple and easily understood systems – including imaginary systems – and works out the pattern of dependence of their attractors on their control parameters. One then looks for these patterns in more complex systems. Very often they fit: the attractors and the relations between changes in the control parameters and changes in the set of attractors that hold for simple systems fit the data for complex systems.

The deeper topological reasons for this are, I gather, now beginning to appear. And so we may soon know what the general mathematical constraints are. (Some of the patterns may be just accidents. Others, for example those involving Feigenbaum's constant, surely are not.)

[1]See [Poston 1978]. [Arnold 1984], is very eloquent about the wildness of wild applications of the theory; and [Zeeman 1976] is very stimulating about the line between explanatory and non-explanatory uses of it. In this connection see also [Morton 1990].

But the general phenomenon is now well-established: by working with simple systems one can develop a zoo of attractors, transitions, and routes to chaos, into the cages of which the behaviour of complex systems fits.[2]

Whether or not I am stating all this absolutely right, the point I am making here is certainly generally right, and important. Call it the Central Point. Its importance is that we can begin to see why a certain class of strategies or stances towards classification of physical systems becomes not a mere heuristic but a source of real explanations referring to the physical sources of their behaviour. What we need – and what I prophesy – is a digesting of these results together with some statistical insight to give a description of strategies which will reasonably often give the right taxonomy of unknown complex systems. (The possibility is of immense epistemological insight. It would be sad if philosophers yet again missed the boat on something so close to their central concerns.)

One importance of the existence of mathematical foundations for explanatory strategies in qualitative physics is that we can see a reason (not the only reason) why one can learn a lot from a mathematical model of a situation. The model will leave out a lot of the detail of the actual situation, in effect reducing it to a much simpler, perhaps physically impossible, situation. But the abstract constraints will restrict the behaviour of the model system to one of the permissible patterns, which may be one exhibited by the actual system. (And one can tell usually whether it is.) So these patterns emerge. As they might not have without a model, either because applying the constraints to the actual case may be beyond us, or, more interestingly, because we do not yet know what the constraints are. There is a potential connection with artificial intelligence issues here, though it is a bit early to wave it about. Workers in AI – for example Smolensky – often have an interest in trying to abstract general complexity-handling techniques from (bits of) physics, for example the study of spin glasses [Hinton et al. 1986], [Smolensky 1986].

[2]For the physics see [Schuster 1984], and [Crutchfield 1986], For philosophical reactions [Hunt 1987], and [Morton 1981]. And for a very readable popular account [Gleick 1987].

The physics in question is usually a matter of annealing or equilibrium-finding: that is, predicting as the energy of a system changes what states it will settle into. That is, of finding its attractors. So one might well suspect links with chaology. Do they exist in any useful form? We don't know yet.

## 13.3 Can't Kant

Taking that as read, two further points that Smith ties to the Central Point must be separated from it. Both are Kantian twists on the Central Point, in different ways. The first is the transcendental turn; it tries to turn the Central Point into an argument for a kind of idealism. You can see the appeal of this since if there are *a priori* constraints on the phenomena which a set of laws can produce, our theorising about the objective world is constrained by something other than the ideal of objective independent fact. Well, there are these constraints, but they don't seem to be of the right kind. One is still allowed to formulate any theory one likes and any theory at all can be true. There are no constraints on the facts but just on what kinds of phenomena can result from them. These constraints are in the end of a piece with the constraint that if you postulate that every pair of lesser slug-eating pine-warblers lays 3 eggs then you must expect that if there is no mortality then when the eggs hatch there will be a family of at least 5 lesser slug-eating pine-warblers attached to each nest. But more interesting.

(Digression: perhaps what we have here is not Smith's misassimilation of physics to the Kantian cause but Kant's misunderstanding of what lies behind his own convictions. I mean: perhaps the real source of *a priori* constraints on our understanding of the world lies not in the fact that the world is not objective and independent but in some subtle and powerful mathematical facts. Smith's appeal to Noether's theorem is suggestive here, because its content does seem to resemble the "first analogy" in the *Critique of Pure Reason* where Kant is arguing, if that is the word, for something like the inevitability of conservation laws. I might understand this better if I could see similarities between Noether's theorem and the results about catastrophes and chaos that Smith and Petitot appeal to. But I can't; it doesn't seem to belong in the same box).

The other elaboration of the Central Point that Smith is attached to is one that does also get my loyalty, but as a conjecture; something that might well be true and would be very interesting if it was. The conjecture also has a Kantian flavour. It is that the constraints postulated by the Central Point may be built into human psychology. (This is like the psychological interpretation of Kant, used freely by all teachers introducing the first critique to students, before hastily repudiating it in favour of the transcendental interpretation. I think the psychological interpretation is the one more likely to be true). I think Smith takes the most plausible line here, by focusing on the connections between catastrophe theory and our ordinary concept of a physical object – a discontinuity against a background of continuous change. For there is quite a lot of evidence – for example in the work of Janet Spelke at Cornell [Spelke 1988] – that human beings conceptualise the world in terms of discrete physical objects from such an early age that the overwhelmingly natural suggestion is that a predisposition to such conceptualisations is built into us. (Note how much more cautiously and flexibly I have put it than saying that the concept of an object is innate.)

Still, there are a lot of gaps here. *Is* there a connection between the discontinuites systematised by catastrophe theory and the human tendency to see a world of objects? Two possible connections come to mind. One would be a sort of justification or vindication of common sense: Darwin was right to build objects into our thinking, because that is in fact the way the world works. The other would make a larger claim about the form of our cognition: seeing the world in terms of discrete objects is part of a very general strategy for coming to grips with complex systems, which forms part of human psychology and is backed up by the mysterious theorems I have alluded to. Well, I suppose I would subscribe to both of these, too. But they are really very vague and dicey, desperately in need of disambiguation and argument. Let me express two serious worries.

The first worry is whether the concept of an object wired into human cognition is the same as anything coming out of qualitative mechanics. After all, it just is not the case that whenever you have a physical object which is intuitively a single coherent object, with its own definite boundaries and its own temporal career through space, you have a catastrophe. You have a singularity of the characteristic function locating the object, but singularities don't mean catastrophes. (And the singularities that the theory standardly applies to are not the boundaries of objects, but the values of quantities associated with their behaviour.) So to see a *deep* (or, really, a halfway informative) connection here you have to say more. I think what you really have to do is to retreat to the second possibility, and see the discrimination of objects as part of a general cognitive strategy that turns out to be qualitative mechanics in disguise.

The second worry is a worry about this second possibility and is in effect the strong AI/weak AI distinction. Suppose that human beings do instantiate a strategy for extracting good regularities out of the flux of information surrounding them which in some sense does the same job as an explanatory strategy in physics. So there are two processes that roughly do the same function: do they instantiate the same algorithm for doing it? That is to push the claim a lot further: if the already rather over-the-top claim is true then we have a hope that algorithms used for getting qualitative solutions for the states physical systems settle into may be a useful tool for simulating some bits of human cognition; but that does not mean that they will be the algorithms that we use.

## 13.4 Quibbles

Before going off in a different direction, some words of warning. First of all neither Smith nor I would deny that many naive physical explanations are wrong, nor in fact that, as is well known, there seems to be a natural human tendency to produce false explanations of some dynamical phenomena: we seem to be predisposed to side with Aristotle against Newton where Newton is right.[3] Since Smith generally sees the echos of Aristotle in some new physics with approval, there is more for him to say here. I imagine his line would be that the Aristotelian tendency is generally reliable – it usually comes up with a good explanation – but at the price of getting some things wrong. So the question is whether this can be said in a way that doesn't trivialise it. I

---

[3]See [McCloskey 1983]

don't myself think one can say anything helpful here without acknowledging that one is dealing with explanation-giving or model-building strategies, and not with theories.

Nor would Smith deny that the range of possible vernacular physicses is very wide. Different cultures or different mental sets can produce very different equally natural conceptualisations of the same phenomena. So innate quasi-*a priori* constraints have to result not in a privileged basic physical theory but in a natural range of theories. But characterising this range is not at all trivial.

## 13.5 Folk Psychology

This last remark brings me on to the next thing I wanted to discuss: the links with folk psychology. There too we have a range of "theories" embedded in different cultures, with a general suspicion that the range has definite limits, so that there are universal constraints on the human conception of the human. There too we have serious doubts whether the natural human conception may not be deeply mistaken so that we ought to face up to the possibility that something built into us or our culture is an obstacle to understanding the truth. (That is the line associated with the "eliminative materialism" originating from Paul Feyerabend and now waved around wildly by Paul and Patricia Churchland, [Churchland 1979], [Churchland 1986].) One disanalogy, though, is that there is not a "naive psychology" subdiscipline of AI. One reason for this, I suspect, is that social psychology is full of interesting generalisations – many of them true! – about the ways people categorise and explain one another. So some of the job has already been done, and there is an imposing body of data which you have to come to terms with if you want to venture into this field.

One interesting confluence of views about folk psychology, now emerging from the work of psychologists and philosophers, gives it two main foci. One focus is on the idea of a physical object: like folk physics. That is, our basic conception of a state of mind is of an attitude towards a particular physical object, either as a perception of it or as a desire for it. There are a lot of problems and controversies here.[4]

[4]For the object-directedness of states of mind see [Morton - to appear]. For rel-

Let me give just one easily-presented bit of evidence, taken from Paul Harris's recent and excellent *Children and Emotion.* Harris is studying children's acquisition of the concept of mind. His data shows clearly that even young babies respond in distinctive ways to expressions of distinct states of mind in those around them. Moreover, these responses are particularly sensitive to the object of the other person's attention. To quote Harris:

> "For instance in one study mothers were asked to express delight, disgust, or neutrality (i.e. silence) towards a particular toy when it was introduced. .. Their children aged 12 months responded appropriately but selectively: they stayed further away from the designated toy and played less with it if the mother expressed disgust, but behaviour towards the other toys and overall mood was unaffected."[5]

The implication is that babies ascribe to their mothers attention to particular objects around them. (Or rather that they do something which is a precursor of psychological ascription and which refers to mothers' attention to things.)

The other focus of folk psychology is on the formation of simulations or models of the state of mind of another. That is, one predicts the behaviour of another in part by putting oneself into a simulacrum, a contained or simplified version, of that person's state. Again it seems that from very early in their socialisation normal human children begin to do this.

There is a tantalising feeling of linkage between this consensus about folk psychology and the Smith-Petitot-Thom picture of naive physics. For not only is the concept of an object fundamental to both, but explanation-by-modelling appears as a basic process in both. It is tempting to put it this way: human beings confronted with the behaviour of other human beings are confronted with a perfect example of the difficulty of predicting the behaviour of a quirky erratic complex system. It is natural to think that human evolution and human cultures have worked out what in the late 20th century we think of as the

---

evant empirical work see [Heelas et al. 1981] On imagination and mental modelling of minds see [Morton 1980] and [Gordon 1988].

[5][Harris 1989].

right way of handling this problem, namely by extracting qualitative predictions from an abstract model of the system. And since the problem is superimposed on the problem of extracting some sense out of the physical world, one might expect the model-building strategy to be a reduplication of the one that makes sense of the world. In effect, then, we not only instantiate an abstract and sophisticated strategy for finding regularities in the world around us, we attribute the strategy to one another. We model one another as modellers. Thus it would be that folk psychology centres on object-directed states and on simulation.

"Tantalising", "tempting", "natural to think", "might expect": that is not what I would call a real argument. But it does suggest directions in which we might look to find something more solid.

## 13.6 What is to be Done?

Everything that I am saying, and everything that Smith is saying, comes to nothing if the promise of qualitative mechanics cannot be made more explicit. That is, there is a very basic task here for philosophers of science, namely to gather together the bits of topology and algebra that underpin the *a priori* constraints on the behaviour of complex systems and systematise them in a useful form. (The mathematics is not all in yet. That is alright: one wants to know what needs proving as well as what is already proved.) If this can be done, then something of enormous epistemological importance will have been accomplished. And something of major interest to workers in AI, too. For the gap between simulating input-output relations typical of human performance and simulating the algorithms which get humans from input to output will have been somewhat lessened.

*If* this can be done. But I suspect that in some measure it can be, and that the result could be one of the intellectual adventures of the next decade.

## References

Arnold, V., [1984]: *Catastrophe Theory*, Springer.

Churchland, Patricia, [1986]: *Neurophysiology*, 430, MIT Press.

Churchland, Paul, [1979]: *Scientific Realism and the Plasticity of Mind*, Cambridge U.P.

Crutchfield, J.P., Farmer, J.D., Packard, N.H., and Shaw, R.S., [1986]: "Chaos" *Scientific American* 255, 46 - 57.

Gleick, J., [1987]: *Chaos*, Penguin.

Gordon, R., [1988]: *The Structure of the Emotions*, Cambridge U.P.

Harris, P., [1989]: *Children and Emotion*, Blackwell.

Heelas, P., and Lock, A., (eds.), [1981]: *Indigenous Psychologies*, Academic Press.

Hinton, G.E., and Sejnowski, T.J., [1986]: "Learning and relearning in Boltzmann machines" in McClelland, J., Rumelhart, D., (eds.) *Parallel Distributed Processing, vol. 1*, MIT Press. 282 - 317.

Hunt, G.M.K., [1987]: "Determinism, Probability, and Chaos" *Analysis* 47, 129 - 133.

McCloskey, [1983]: "Intuitive Physics", *Scientific American*, 248, 122-130

Morton, A., [1980]: *Frames of Mind*, Ch.3. Oxford U.P.

Morton, A., [1981]: "The Chaology of Mind", *Analysis* 48, 135-142.

Morton, A., [to appear 1990]: "Mathematical modelling and contrastive explanation", *Canadian Journal of Philosophy.*

Morton, A., [to appear]: "The inevitability of folk psychology" in Radu Bogdan, (ed.) *Mind and Common Sense*, Cambridge U.P.

Poston, T., and Stewart, I., [1978]: *Catastrophe Theory and its Applications*, Pittman.

Schuster, H.G., [1984]: *Deterministic Chaos*, Physik-Verlag.

Smolensky, P., [1986]: "Information processing in dynamic systems: foundations of harmony theory", in McClelland, J., Rumelhart, D., (eds.) *Parallel Distributed Processing, vol 1*, MIT Press.

Spelke, E.S., [1988]: "The origins of physical knowledge", in Wiskrantz, L., (ed.) *Thought without language*, Oxford U.P.

Zeeman, C., [1976]: "Catastrophe Theory", *Scientific American.*

# Bibliographic Index